T0184879

Springer Monographs in Mathematics

More information about this series at http://www.springer.com/series/3733

Stanley Eigen • Arshag Hajian • Yuji Ito
Vidhu Prasad

Weakly Wandering Sequences in Ergodic Theory

 Springer

Stanley Eigen
Department of Mathematics
Northeastern University
Boston, MA, USA

Arshag Hajian
Department of Mathematics
Northeastern University
Boston, MA, USA

Yuji Ito (emeritus)
Department of Mathematics
Keio University
Yokohama, Japan

Vidhu Prasad
Department of Mathematical Sciences
University of Massachusetts Lowell
Lowell, MA, USA

ISSN 1439-7382   ISSN 2196-9922 (electronic)
ISBN 978-4-431-56400-3   ISBN 978-4-431-55108-9 (eBook)
DOI 10.1007/978-4-431-55108-9
Springer Tokyo Heidelberg New York Dordrecht London

Mathematics Subject Classification (2010): 37A40, 37A45, 28D05, 11B13, 11B05

Printed on acid-free paper

Springer is part of Springer Science+Business Media (www.springer.com)

*This monograph is dedicated to
Professor Shizuo Kakutani 1911–2004*

Foreword

Weakly wandering (*ww*) sets made their first appearance over 50 years ago in [23, 39] (see also [32, 40]). The late Professor Shizuo Kakutani of Yale University was instrumental in advising and directing us in the development of the above works. Initially, the appearance of *ww* sets and sequences was a surprising event, yet at the same time quite useful in the study of problems connected with the existence of finite invariant measures. Soon it was realized that *ww* sequences were always present for all ergodic transformations that did not preserve a finite measure. Professor Kakutani felt that this was an important fact and strongly encouraged us to study the role of these sequences in the classification of infinite ergodic transformations. During the years that followed, we would meet periodically with him in the southern New England region from New Haven, to Providence, Boston, and Amherst, spend long periods studying various problems in ergodic theory, and often discuss questions connected with properties of *ww* sequences. He was aware that the *ww* and related sequences associated with infinite ergodic transformations were powerful isomorphism invariants, and he urged us to investigate their properties. More than anyone else he had a keen sense of understanding the nature of infinite measure spaces and properties of the transformations defined on them.

During the following 40 years we continued our joint work and published several articles on the properties of infinite ergodic transformations. It was during one of our frequent meetings that Professor Kakutani suggested the writing of a monograph which gathered most of the published and unpublished results that we had obtained. We had just started on that project when it was interrupted by his untimely departure.

Professor Kakutani was a constant force guiding and encouraging us to continue working and looking into the effect of *ww* and related sequences on the behavior of infinite ergodic transformations. This monograph is a result of that. Eliminating the contribution of any of the co-authors from this monograph would make it noticeably weaker. On the other hand, without Professor Kakutani's contribution and constant encouragement this monograph would not exist.

Preface

The material in this monograph is self-contained. A basic knowledge of measure theory as taught to beginning graduate students is the only prerequisite needed to read and understand the material presented. Prior knowledge of ergodic theory is useful but not necessary. Some fundamental properties of ergodic transformations preserving a σ-finite infinite measure as discussed in Chap. 3 follow easily from Birkhoff's Individual Ergodic Theorem. However, even these properties are developed and proven directly.

In Chap. 1 we discuss in some detail various conditions for the existence of a finite invariant measure. In 1932, E. Hopf [37] presented an interesting geometric condition that was necessary and sufficient for the existence of a finite invariant measure for a measurable and nonsingular transformation. Later in 1956, Y. Dowker [8] discussed the same problem and presented a different condition involving the measure of iterates of the images of measurable sets. Initially, the two conditions, the one presented by Hopf (**H**) and the other by Dowker (**D1**), did not seem to be obviously related except for the fact that they were both necessary and sufficient conditions for the solution of the same problem. The attempt to prove their equivalence by direct arguments on the other hand revealed the interesting and unexpected fact that all infinite ergodic transformations possessed weakly wandering (*ww*) sets: these are sets of positive measure with an infinite number of mutually disjoint images under a sequence of integers called a *ww* sequence. We also mention some minor facts that emerged during our attempt to show by direct arguments the equivalence of the various conditions of the Finite Invariant Measure Theorem 1.2.1. One of these is Proposition (subadditive) 1.2.2, which is a slight generalization related to a well-known result on the equivalence of finite measures, and another is Proposition (additive) 1.2.4, which exhibits the additive nature of the Cesaro sums of the measure of iterates of measurable sets. We also point out that condition (**W***), which is a (seemingly) stronger condition than condition (**W**), happens to be equivalent to it. The following interesting-sounding remark is a consequence of Theorems 1.1.3 and 1.2.1: a simple strengthening of the statement

A measurable transformation is recurrent if and only if it does not possess wandering sets

is

A measurable transformation is strongly recurrent if and only if it does not possess weakly
wandering sets,

and both statements are true.

In Chap. 2 we discuss properties of transformations that do not possess a finite
invariant measure. While writing this monograph we were often tempted to relabel
this chapter: "The Non-Existence of a Finite Invariant Measure." The appearance
of *ww* sequences for such transformations turned out to be a powerful tool in the
classification of infinite ergodic transformations. In time the existence of these
sequences implied the existence of even more interesting sequences connected
with ergodic transformations without finite invariant measure. One such was
the appearance of an equally unexpected sequence that we called an exhaustive
weakly wandering (*eww*) sequence: this is a *ww* sequence where the images of
a corresponding *ww* set cover the whole space. In Definition 2.1.1 we introduce
a more complicated sequence for measurable transformations, which we call a
strongly weakly wandering (*sww*) sequence. At first encounter it seems that *sww*
sequences have unnecessarily complicated properties and are difficult to understand.
However, for ergodic transformations without a finite invariant measure we are able
to show without too much effort the existence of *sww* sequences, and these in turn
imply the existence of a special kind of *eww* sequence. Moreover, we do not know a
better way of proving the existence of *eww* sequences for transformations without a
finite invariant measure. Initially, after we were confronted with the existence of *ww*
and *eww* sequences, for a very long time we were not aware of any general property
of a *ww* sequence that insured its possessing an *eww* subsequence. To our surprise
Proposition 2.2.4 accomplishes that.

In Chap. 3 we discuss infinite ergodic transformations and mention the following
important property that these transformations possess: for any two sets A and
B of finite measure $\lim_{n\to\infty} \frac{1}{n} \sum_{i=0}^{n-1} m(T^i A \cap B) = 0$. This property follows
immediately from Birkhoff's Individual Ergodic Theorem. However, we present
a direct and elementary proof of it. Next we introduce *recurrent* sequences for
infinite ergodic transformations. These are sequences of integers that have a finite
intersection with every *ww* sequence for an infinite ergodic transformation. As we
shall see in Chap. 4 there exist infinite ergodic transformations that possess recurrent
sequences and others that do not. We discuss both classes of infinite ergodic transformations. The infinite ergodic transformations that possess recurrent sequences are
interesting in connection with the various conditions discussed in Theorem 1.2.1
(the Finite Invariant Measure Theorem). The infinite ergodic transformations that
do not possess recurrent sequences happen to be even more interesting. For such
transformations we prove Theorems 3.3.11 and 3.3.12 where we show that these
transformations possess *ww* and *eww* growth sequences. A consequence of this
is the fact that for infinite ergodic transformations without recurrent sequences
every infinite sequence of integers contains an *eww* subsequence. Incidentally, after

working with the presence and absence of recurrent sequences and discussing their connection with *ww* and *eww* sequences, we urge the reader to refrain from naming any feature of an infinite ergodic transformation as being some sort of mixing. As tempting as it may seem, we feel that labeling any property of an infinite ergodic transformation as some type of "mixing" is wrong.

In Chap. 4 we present three important and basic examples of infinite ergodic transformations. The First Basic Example was constructed soon after it was realized that a consequence of the main theorem of [32] was the fact that every infinite ergodic transformation possesses *ww* sets. Initially, this fact was a bit difficult to digest. In particular, among all the existing examples of infinite ergodic transformations we could not exhibit a single *ww* set. This prompted us to construct the First Basic Example in [33]. Our object was to see a concrete *ww* set and the *ww* sequence associated with it. Employing the machinery of induced transformations we succeeded in constructing the desired transformation together with the *ww* sequence and set. Subsequently, we were compensated with a few extra and unexpected rewards. The *ww* set that we had constructed happened to be, in fact, an *eww* set as well. Up to that point the existence of an *eww* set for any infinite ergodic transformation was unthinkable. Our surprise was even greater when we noticed that this *eww* set was actually a set of finite measure. A number of years later the Second Basic Example was constructed in [31] with the purpose of showing that the existence of an *eww* set of finite measure was not shared by all infinite ergodic transformations. This was accomplished by showing that the commutators of the Second Basic Example contained a non-measure-preserving transformation. This fact was used in showing how to construct in a systematic way an ergodic transformation that does not preserve any σ-finite invariant measure. The Third Basic Example that we present next was actually constructed well before the two preceding examples. It was originally constructed in [23] to show that the further weakening of condition (**D3**) of Theorem 1.2.1 of Chap. 1 as a necessary and sufficient condition for the existence of a finite invariant measure was not possible. Later it was also realized that, unlike the previous two examples, this basic example was an infinite ergodic transformation that did not possess recurrent sequences. Finally, a variant of an example discussed by E. Hopf in his book [38] is sketched as another example of an infinite ergodic transformation that does not possess recurrent sequences. This example can also be regarded as a simple realization of symmetric random walk on the integers as an infinite ergodic transformation.

In Chap. 5 we consider various collections of infinite subsets of the integers $\mathbb{Z}$ associated to an infinite ergodic transformation, and discuss a number of properties of these collections. In particular we give descriptions of collections of *ww*, recurrent, and dissipative sequences for the transformation T. We end the chapter with a topological description of these and other collections in terms of the Stone–Čech compactification $\beta\mathbb{Z}$ of $\mathbb{Z}$. Theorem 5.1.2 of this chapter is a particularly interesting theorem concerning the behavior of transformations that possess an *eww* set of finite measure.

In Chap. 6 we examine various isomorphism invariants for infinite ergodic transformations. We begin with *eww* sequences and note that an isomorphism, besides leaving an *eww* sequence invariant, must also map *eww* sets to each other. We then show by example that two such sets for a common sequence may sit quite differently within a transformation. We then introduce the α-type of an ergodic transformation as an isomorphism invariant and show its relation to the recurrent sequences of the transformation. In the second part of the chapter we examine a class of transformations for which a complete characterization of the recurrent sequences can be described. We end the chapter with a result on how the growth rate of the *ww* sequences for a transformation is also an isomorphism invariant.

In Chap. 7 we show that *eww* sequences are related to complementing pairs of subsets of integers which tile the set of integers. We begin with a review of known results for tilings of the integers and point out that the tools used when one member of a pair is finite are not applicable when both members are infinite. We then show how such tilings of the integers arise in ergodic theory and use the fact that one member is the hitting times of a generic point to an *eww* set to obtain a characterization of *eww* sequences. A number of examples are given which indicate the difficulties of the subject. Finally, we conclude the chapter by showing how *p*-adic analysis is related to *eww* sequences for some infinite ergodic transformations.

Boston, MA, USA   Stanley Eigen
Boston, MA, USA   Arshag Hajian
Yokohama, Japan   Yuji Ito
Lowell, MA, USA   Vidhu Prasad

Contents

Chapter 1
Existence of Finite Invariant Measure

In this chapter we discuss properties of a transformation T that are equivalent to the transformation being recurrent. We show that strengthened versions of these properties, together with a few more properties of T, are necessary and sufficient conditions for the existence of a finite invariant measure μ for T.

We consider transformations T that are invertible (1-1, onto) maps defined on a σ-finite Lebesgue measure space $(X, \mathscr{B}, m)$. Even when not mentioned explicitly, all the transformations we consider are assumed to be measurable ($A \in \mathscr{B}$ if and only if $TA \in \mathscr{B}$) and nonsingular ($m(A) = 0$ if and only if $m(TA) = 0$). Throughout this monograph all the sets we mention are assumed to be measurable, and often we make statements ignoring sets of measure 0.

We say that m is an invariant measure for a transformation T if $m(TA) = m(A)$ for all $A \in \mathscr{B}$. Two measures m and μ are *equivalent* ($m \sim \mu$) if m and μ have the same sets of measure zero. When an invariant measure $\mu \sim m$ exists for T we say that T preserves the measure μ, or T is a measure-preserving transformation.

In this section we study necessary and sufficient conditions for the existence of a finite T-invariant measure $\mu \sim m$.

We remark that since $(X, \mathscr{B}, m)$ is a σ-finite measure space, it is always possible to find a finite measure $m' \sim m$. Namely, since $X = \bigcup_{i=1}^{\infty} A_i (disj)$, where A_i are sets of finite positive measure for $i = 1, 2, \ldots$, we define

$$m'(B) = \sum_{i=1}^{\infty} \frac{m(B \cap A_i)}{2^i m(A_i)} \quad \text{for} \quad B \in \mathscr{B}.$$

Therefore, whenever a transformation T is not assumed to be measure-preserving, without loss of generality we may assume that T is defined on a finite measure space $(X, \mathscr{B}, m)$ with $m(X) = 1$.

© Springer Japan 2014
S. Eigen et al., *Weakly Wandering Sequences in Ergodic Theory*,
Springer Monographs in Mathematics, DOI 10.1007/978-4-431-55108-9_1

1.1   Recurrent Transformations

Let us make the following definitions:

Definition 1.1.1. Let T be a measurable and nonsingular transformation defined on the measure space $(X, \mathscr{B}, m)$.

- T is a *recurrent* transformation if
  $m(A) > 0 \implies$ for a.a. $x \in A$ there is an integer $n > 0$ such that $T^n x \in A$.
- A is a *wandering* set for T if
  $m(A) > 0$, and $T^i A \cap T^j A = \emptyset$ for $i, j \in \mathbb{Z}$ and $i \neq j$.
- Two sets A and B are *finitely equivalent*, $A \approx B$, if for some integer $p > 0$
  $A = \bigcup_{i=1}^{p} A_i (disj)$, $B = \bigcup_{i=1}^{p} B_i (disj)$, and for a set of p integers $\{n_i : 1 \leq i \leq p\}$ $T^{n_i} A_i = B_i$ for $1 \leq i \leq p$.
- Two sets A and B are *countably equivalent*, $A \sim B$, if
  $A = \bigcup_{i=1}^{\infty} A_i (disj)$, $B = \bigcup_{i=1}^{\infty} B_i (disj)$, and for a sequence of integers $\{n_i : i \geq 1\}$ $T^{n_i} A_i = B_i$ for $1 \leq i < \infty$.
- A set A is *strongly recurrent* if $\{n : m(T^n A \cap A) > 0\}$ is relatively dense in $\mathbb{Z}$, or equivalently:
  there is an integer $k > 0$ such that $\max_{0 \leq i \leq k} m(T^{n+i} A \cap A) > 0$ for all $n \in \mathbb{Z}$.
- An infinite sequence of integers $\{n_i : i \geq 0\}$ is a *weakly wandering (ww) sequence* for T if there is a set W of positive measure such that
  $T^{n_i} W \cap T^{n_j} W = \emptyset$ for $i, j \geq 0$ and $i \neq j$.
  Often we will say W is a *ww* set (for T) (with the sequence $\{n_i\}$), or $\{n_i\}$ is a *ww* sequence (for T) (with the set W).

The following lemma about wandering sets is used in the proof of the Recurrence Theorem that follows.

Lemma 1.1.2 (Wandering Sets). *The following two conditions for a nonsingular transformation T on $(X, \mathscr{B}, m)$ are equivalent.*

(1) *T does not admit any wandering sets.*
(2) *For a measurable function $f(x)$, if $f(Tx) \leq f(x)$ a.e., then $f(Tx) = f(x)$ a.e.*

Proof. **(1)** $\Rightarrow$ **(2)**: Assume condition **(2)** does not hold. Then there is a measurable function $f(x)$ so that $f(Tx) \leq f(x)$ a.e. and $m\{x : f(Tx) < f(x) < \infty\} > 0$. Therefore there exists a constant c such that if $W = \{x : f(Tx) \leq c < f(x)\}$ then $m(W) > 0$.
For $x \in W$ we have: $f(T^n x) \leq f(T^{n-1} x) \leq \cdots \leq f(Tx) \leq c$.
For $x \in T^{-n} W$, since $T^n x \in W$, we have: $c < f(T^n x)$.
Then $T^n W \cap W = \emptyset$ for all $n > 0$. Thus $T^i W \cap T^j W = T^j (T^{i-j} W \cap W) = \emptyset$ for $i > j$. In other words, W is a wandering set for T. This is a contradiction to **(1)**.
(2) $\Rightarrow$ **(1)**: Assume condition **(1)** does not hold.
Let W be a wandering set for T, and let $W^* = \bigcup_{n=0}^{\infty} T^{-n} W$.

Let $f(x) = \chi_{W^*}(x)$ be the characteristic function of the set W^*. Then $f(Tx) = \chi_{W^*}(Tx) = \chi_{T^{-1}W^*}(x) \leq f(x)$, and for $x \in W$, $0 = f(Tx) < f(x) = 1$.

This is a contradiction to (2). □

The following theorem considers conditions which are equivalent for a transformation T being recurrent.

Theorem 1.1.3 (Recurrence). *Let T be a measurable and nonsingular transformation on $(X, \mathscr{B}, m)$. Then the following conditions are equivalent:*

(1R) $m(A) > 0 \implies$ *for a.a.* $x \in A$ *there is an integer* $n > 0$ *such that* $T^n x \in A$ *(T is recurrent).*

(2R) $m(A) > 0 \implies$ *for a.a.* $x \in A$ *there are infinitely many integers* $n > 0$ *such that* $T^n x \in A$.

(3R) *For a measurable function* $f(x) > 0$ *we have:* $\sum_{n=1}^{\infty} f(T^n x) = \infty$ *a.e.*

(4R) $m(A) > 0 \implies$ *there is an integer* $n > 0$ *such that* $m(T^n A \cap A) > 0$.

(5R) *T does not admit any wandering sets.*

(6R) *If A is finitely equivalent to B $(A \approx B)$, $m(A) > 0$ and $A \supset B$, then* $m(A \smallsetminus B) = 0$.

Proof.

(1R) $\Rightarrow$ (2R): Repeated application of condition **(1R)** implies **(2R)**.

(2R) $\Rightarrow$ (3R): Assume condition **(3R)** does not hold. Then there is a measurable function f, $f(x) > 0$, and a set A, $m(A) > 0$, such that for $x \in A$ we have $\sum_{n=1}^{\infty} f(T^n x) < \infty$. For some $\varepsilon > 0$, by possibly removing a small subset of A, we may assume that $m(A) > 0$ and $f(x) \geq \varepsilon > 0$ for $x \in A$.

For $p \geq 1$ let $A_p = \{x \in A : \sum_{n=1}^{\infty} f(T^n x) \leq p\}$. Then $A = \bigcup_{p=1}^{\infty} A_p$, and $m(A) > 0$ implies that for some $p > 0$ we have $m(A_p) > 0$. Then $f(x) \geq \varepsilon > 0$ for $x \in A_p$, and $\sum_{n=1}^{\infty} f(T^n x) \leq p$. In other words, for $x \in A_p$, the cardinality of $\{n : T^n x \in A_p\}$ is not greater than p/ε, a finite number. This is a contradiction to condition **(2R)**.

(3R) $\Rightarrow$ (4R): Assume condition **(4R)** does not hold. Then there is a set A of positive measure such that $T^n A \cap A = \emptyset$ for all $n > 0$. We let

$$
f(x) = \begin{cases} 1 & \text{if } x \in X \smallsetminus \bigcup_{i=1}^{\infty} T^i A, \\ 1/2^n & \text{if } x \in T^n A \text{ for } n > 0. \end{cases}
$$

Then $f(x) > 0$, and for $x \in A$ we have

$$
\sum_{n=0}^{\infty} f(T^n x) = \sum_{n=0}^{\infty} 1/2^n < \infty.
$$

This is a contradiction to condition **(3R)**.

(4R) $\Rightarrow$ (5R):   If condition (5R) does not hold then the existence of a wandering set contradicts condition (4R).

(5R) $\Rightarrow$ (6R):   Suppose there are sets A, B with $A \supset B$ and for some $p > 0$ there are integers $\{n_i : 1 \le i \le p\}$ such that

$$A = \bigcup_{i=1}^{p} A_i\,(disj), \quad B = \bigcup_{i=1}^{p} B_i\,(disj) \quad \text{and} \quad A_i = T^{n_i} B_i \quad \text{for } 1 \le i \le p.$$

We may assume $n_i \ne 0$ for each i. Let $I_1 = \{i : 1 \le i \le p, n_i > 0\}$ and $I_2 = \{i : 1 \le i \le p, n_i < 0\}$. We define

$$f(x) = \sum_{i \in I_1} \sum_{j=0}^{n_i-1} \chi_{A_i}(T^j x) - \sum_{i \in I_2} \sum_{j=n_i}^{-1} \chi_{A_i}(T^j x).$$

Since $\chi_{A_i}(T^{n_i} x) = \chi_{B_i}(x)$, for each i we obtain

$$f(Tx) = f(x) - \sum_{i \in I_1}\left(\chi_{A_i}(x) - \chi_{B_i}(x) \right) + \sum_{i \in I_2}\left(\chi_{B_i}(x) - \chi_{A_i}(x) \right)$$

$$= f(x) - \left(\chi_A(x) - \chi_B(x) \right).$$

Therefore $f(x) - f(Tx)$ is equal to the characteristic function $\chi_{A \setminus B}$ which is nonnegative a.e. and takes the value 1 on $A \setminus B$. Lemma 1.1.2 then implies $m(A \setminus B) = 0$, and this proves condition (6R).

(6R) $\Rightarrow$ (1R):   Assume condition (1R) does not hold. Then there is a set C such that for $x \in C$ we have $T^n x \notin C$ for any $n > 0$. Let

$$A = \bigcup_{n=0}^{\infty} T^n C \quad \text{and} \quad B = \bigcup_{n=1}^{\infty} T^n C.$$

Then $m(A \setminus B) = m(C) > 0$ and $TA = B$. This contradicts condition (6R).

$\square$

1.2   Finite Invariant Measure

Suppose T admits a wandering set W with $m(W) > 0$. Then for any measure $\mu \sim m$ it is clear that $\mu(W) > 0$. If μ is an invariant measure for T then the infinite number of mutually disjoint images of W under T will have the same positive μ measure. It follows that the condition: T does not admit wandering sets, namely condition (5R) of Theorem 1.1.3, is a necessary condition for the existence of a finite invariant measure $\mu \sim m$. In view of the Recurrence Theorem any one of the equivalent conditions **(1R)–(6R)** of Theorem 1.1.3 is also a necessary condition for the existence of a finite T-invariant measure $\mu \sim m$. However, as we shall see later, the converse is not true; in other words, none of the conditions of the Recurrence Theorem is a sufficient condition for the existence of a finite T-invariant measure $\mu \sim m$. To obtain sufficient conditions for the existence of an invariant measure μ the conditions in the Recurrence Theorem need to be strengthened.

We list below conditions on a transformation T that various authors introduced and showed to be necessary and sufficient for the existence of a finite invariant measure $\mu \sim m$.

In [8] the following condition was introduced:

(D1): $m(A) > 0 \implies \liminf\limits_{n \to \infty} m(T^n A) > 0.$

In [2] the following weakened version of condition **(D1)** was introduced:

(C): $m(A) > 0 \implies \liminf\limits_{n \to \infty} \dfrac{1}{n} \sum\limits_{i=0}^{n-1} m(T^i A) > 0.$

In [9] the following weakened version of condition **(C)** was introduced:

(D2): $m(A) > 0 \implies \limsup\limits_{n \to \infty} \dfrac{1}{n} \sum\limits_{i=0}^{n-1} m(T^i A) > 0.$

In [26] the following strengthened version of condition **(3R)** was introduced:

(F): $f(x) > 0 \text{ measurable} \implies \sum\limits_{i=1}^{\infty} f(T^{n_i} x) = \infty$ a.e. for any sequence $\{n_i\}$.

In [24] the following strengthened version of condition **(4R)** was introduced:

(S): $m(A) > 0 \implies A$ is strongly recurrent.

In [32] the following strengthened version of condition **(5R)** was introduced:

(W): $m(A) > 0 \implies A$ is not weakly wandering.

In [37] the following strengthened version of condition **(6R)** was introduced:

(H): A countably equivalent to B $(A \sim B), m(A) > 0, A \supset B \implies m(A \setminus B) = 0.$

In trying to find an invariant measure $\mu \sim m$ for T, it is useful to study the asymptotic behavior of the measure of the iterates $m(T^n A)$ of a measurable set A. Actually, the conditions **(D1)**, **(C)** and **(D2)**, which are necessary and sufficient conditions for the existence of a finite invariant measure, refer to the behavior of the values of $\{m(T^n A)\}$ or their Cesaro sums. These conditions are closer in nature to the behavior of an invariant measure that is being sought. The other conditions, namely **(F)**, **(S)**, **(W)**, **(H)**, which are strengthened versions of the corresponding conditions **(R3)**, **(R4)**, **(R5)**, **(R6)** respectively, of the Recurrence Theorem 1.1.3 are more geometric in nature. They depend on the class of measures equivalent to m rather than on the measure of the iterates $\{T^n A\}$. In the following theorem we show by direct arguments that all of the above conditions are equivalent and are necessary and sufficient conditions for the existence of a finite invariant measure $\mu \sim m$ for T. The results and methods used in proving Theorem 1.2.1 were discussed in detail and used in [24, 26, 32].

Theorem 1.2.1 (Finite Invariant Measure). *Let T be a measurable and non-singular transformation defined on the finite measure space $(X, \mathscr{B}, m)$. Then conditions **(D1)**, **(C)**, **(D2)**, **(F)**, **(S)**, **(W)**, **(H)** mentioned above are equivalent to each other. Furthermore, they are necessary and sufficient conditions for the existence of a finite invariant measure $\mu \sim m$ for T.*

To prepare the way for the proof of the Finite Invariant Measure Theorem, we first recall some definitions about set functions and prove several propositions.

Let $(X, \mathscr{B})$ be a measurable space, and let λ be a real-valued nonnegative set function defined on $\mathscr{B}$.

λ is said to be *monotonic* if
$\quad \lambda(A) \leq \lambda(B)$ for any two sets A, B with $A \subset B$.
λ is said to be *subadditive* if
$\quad \lambda(A \cup B) \leq \lambda(A) + \lambda(B)$ for any two sets A, B.
λ is said to be *superadditive* if
$\quad \lambda(A \cup B) \geq \lambda(A) + \lambda(B)$ for any two disjoint sets A, B.
μ is said to be *absolutely continuous* with respect to λ if
$\quad \lambda(A) = 0 \implies \mu(A) = 0$.
μ is said to be *uniformly absolutely continuous* with respect to λ if
$\quad$ for any $\varepsilon > 0$ there exists a $\delta > 0$ such that $\lambda(A) < \delta \implies \mu(A) < \varepsilon$.

We observe that if λ is nonnegative and superadditive then λ is monotonic. For any two set functions λ and μ defined on $\mathscr{B}$ it is clear that uniform absolute continuity implies absolute continuity, while the converse is not always true. It is known, however, that the converse is true if both λ and μ are finite measures defined on $\mathscr{B}$. The following proposition may be considered a generalization of this fact.

Proposition 1.2.2 (Subadditive). *For the finite measure space $(X, \mathscr{B}, m)$ let λ be a real-valued, nonnegative, monotonic and subadditive set function defined on $\mathscr{B}$. If m is absolutely continuous with respect to λ, then m is uniformly absolutely continuous with respect to λ.*

Proof. We assume Proposition 1.2.2 does not hold. Then there is an $\varepsilon > 0$ and a sequence of subsets $\{B_n : n = 1, 2, \ldots\}$ such that $\lambda(B_n) < 1/2^n$, and $m(B_n) \geq \varepsilon$ for $n \geq 1$. For each $n = 1, 2, \ldots$ let p_n be a positive integer such that $p_n > n$ and

$$m(\Delta_n) < \varepsilon/2^{n+1} \quad \text{where} \quad \Delta_n = \bigcup_{k=n}^{\infty} B_k \smallsetminus \bigcup_{k=n}^{p_n} B_k.$$

This is possible since $m(X) < \infty$ by assumption.
Let us put

$$B^{**} = \bigcap_{n=1}^{\infty} \bigcup_{k=n}^{\infty} B_k \left(\equiv \limsup_{n \to \infty} B_n \right) \quad \text{and} \quad B^* = \bigcap_{n=1}^{\infty} \bigcup_{k=n}^{p_n} B_k.$$

Then

$$B^* = \bigcap_{n=1}^{\infty} \left(\bigcup_{k=n}^{\infty} B_k \smallsetminus \Delta_n \right) \supset \bigcap_{n=1}^{\infty} \bigcup_{k=n}^{\infty} B_k \smallsetminus \bigcup_{n=1}^{\infty} \Delta_n = B^{**} \smallsetminus \bigcup_{n=1}^{\infty} \Delta_n \,,$$

and hence

$$m(B^*) \geq m(B^{**}) - \sum_{n=1}^{\infty} m(\Delta_n) > \varepsilon - \sum_{n=1}^{\infty} \frac{\varepsilon}{2^{n+1}} = \frac{\varepsilon}{2} > 0 \,.$$

On the other hand,

$$\lambda(B^*) \leq \lambda \left(\bigcup_{k=n}^{p_n} B_k \right) \leq \sum_{k=n}^{p_n} \lambda(B_k) < \sum_{k=n}^{p_n} \frac{1}{2^k} < \frac{1}{2^{n-1}} \to 0 \ \text{ as } \ n \to \infty;$$

therefore $\lambda(B^*) = 0$, and this is a contradiction. $\square$

The following proposition on superadditive set functions is also useful.

Proposition 1.2.3 (Superadditive). *Let $(X, \mathscr{B})$ be a measurable space. Let λ be a real-valued, nonnegative, superadditive (and hence monotonic) set function defined on $\mathscr{B}$. If $\{B_n : n = 1, 2, \ldots\}$ is a decreasing sequence of measurable sets, then for any $\varepsilon > 0$ there exists a positive integer n_0 such that $\lambda(B_{n_0} \smallsetminus B_n) < \varepsilon$ for any $n > n_0$.*

Proof. From the monotonicity of λ follows that $\lim_{n \to \infty} \lambda(B_n) = \beta \geq 0$ exists and $\lambda(B_n) \geq \beta$ for $n = 1, 2, \ldots$.

For any $\varepsilon > 0$ let n_0 be a positive number such that $\lambda(B_{n_0}) < \beta + \varepsilon$. Then

$$\lambda(B_{n_0} \smallsetminus B_n) + \lambda(B_n) \leq \lambda(B_{n_0}) < \beta + \varepsilon \quad \text{for} \ n > n_0,$$

and hence

$$\lambda(B_{n_0} \smallsetminus B_n) < \lambda(B_{n_0}) - \lambda(B_n) < \varepsilon.$$

$\square$

For a measurable and nonsingular transformation T defined on the finite measure space $(X, \mathscr{B}, m)$ we consider the following set functions:

$$\sigma_n(A) = \frac{1}{n} \sum_{i=0}^{n-1} m(T^i A),$$

$$\overline{\sigma}(A) = \limsup_{n \to \infty} \sigma_n(A) \quad \text{and} \quad \underline{\sigma}(A) = \liminf_{n \to \infty} \sigma_n(A).$$

It is clear that $\overline{\sigma}$ and $\underline{\sigma}$ are real-valued, nonnegative and monotonic set functions defined on $\mathscr{B}$. Both are invariant under T, and $\overline{\sigma}$ is subadditive while $\underline{\sigma}$ is superadditive. The following proposition exhibits the additive nature of both $\overline{\sigma}$ and $\underline{\sigma}$ on finitely equivalent sets.

Proposition 1.2.4 (Additive). *Let T be a measurable and nonsingular transformation defined on the finite measure space $(X, \mathscr{B}, m)$. Let $A_0, A_1, \ldots, A_{r-1}$ be a finite collection of sets which are finitely equivalent with each other. Then*

$$\limsup_{n \to \infty} \sum_{i=0}^{r-1} \sigma_n(A_i) = r\overline{\sigma}(A_0), \tag{1.1}$$

and

$$\liminf_{n \to \infty} \sum_{i=0}^{r-1} \sigma_n(A_i) = r\underline{\sigma}(A_0). \tag{1.2}$$

In particular, if the sets $A_0, A_1, \ldots, A_{r-1}$ are mutually disjoint, then

$$\overline{\sigma}\left(\bigcup_{p=0}^{r-1} A_p \right) = r\overline{\sigma}(A_0), \tag{1.3}$$

and

$$\underline{\sigma}\left(\bigcup_{p=0}^{r-1} A_p \right) = r\underline{\sigma}(A_0). \tag{1.4}$$

Proof. We first observe that for any set C, any integers $p \in \mathbb{Z}$ and $n \in \mathbb{N}$, we have:

$$\left| \sigma_n(C) - \sigma_n(T^p C) \right| = \left| \frac{1}{n} \sum_{k=0}^{n-1} m(T^k C) - \frac{1}{n} \sum_{k=p}^{p+n-1} m(T^k C) \right|$$

$$\leq \frac{2|p|}{n} m(X). \tag{1.5}$$

Suppose $C \approx D$, then for a positive integer s

$$C = \bigcup_{j=1}^{s} C_j \, (disj), \ D = \bigcup_{j=1}^{s} D_j \, (disj) \text{ and } T^{p_j} C_j = D_j \text{ for } 1 \leq j \leq s.$$

From (1.5) follows that for any increasing sequence $n_k \longrightarrow \infty$

$$|\sigma_{n_k}(C) - \sigma_{n_k}(D)| = \left| \sum_{j=1}^{s} [\sigma_{n_k}(C_j) - \sigma_{n_k}(D_j)] \right|$$

$$\leq \sum_{j=1}^{s} \frac{2|p_j|}{n_k} m(X) \longrightarrow 0.$$

From this we see

$$\left| \sum_{i=0}^{r-1} \sigma_{n_k}(A_i) - r\sigma_{n_k}(A_0) \right| \leq \sum_{i=0}^{r-1} \left| \sigma_{n_k}(A_i) - \sigma_{n_k}(A_0) \right| \longrightarrow 0 \quad \text{as} \quad n_k \longrightarrow \infty.$$

$$(1.6)$$

Then (1.1) and (1.2) follow from (1.6), while (1.3) and (1.4) are consequences of (1.1) and (1.2).   □

Proposition 1.2.5. *Let T be a nonsingular transformation defined on the finite measure space $(X, \mathscr{B}, m)$. Let A be a set of positive measure such that*

$$\liminf_{n \to \infty} m(T^n A) = 0. \tag{1.7}$$

Then for $0 < \varepsilon < m(A)$ there is a set $A' \subset A$, with $m(A') < \varepsilon$, such that $B = A \smallsetminus A'$ is not strongly recurrent.

Proof. Let the set A and $\varepsilon > 0$ be given as above, and for $k \geq 1$ let $\varepsilon_k = \varepsilon/(k2^k)$.

For each $k \geq 1$ we choose an integer $n_k > 0$ such that $m\left(T^{n_k-i} A \cap A\right) < \varepsilon_k$ for $0 \leq i \leq k-1$. This is possible since A satisfies (1.7), and since $m(X) < \infty$ the nonsingularity of T implies the absolute continuity of the measures $m_n(E) = m(T^n E)$ with respect to m. Let us put

$$A' = \bigcup_{k=1}^{\infty} \bigcup_{i=0}^{k-1} T^{n_k-i} A \cap A.$$

Then

$$m(A') \leq \sum_{k=1}^{\infty} \sum_{i=0}^{k-1} m(T^{n_k-i} A \cap A) < \sum_{k=1}^{\infty} k\varepsilon_k = \varepsilon.$$

Let $B = A \smallsetminus A'$. It is easy to see that $m(B) \geq m(A) - \varepsilon > 0$, and

$$T^{n_k-i} B \cap B \subset T^{n_k-i} A \cap (A \smallsetminus A') = \emptyset \quad \text{for} \quad 1 \leq i \leq k-1 \quad \text{and} \quad k \geq 1.$$

This says that for each $k \geq 1$ there is an integer $n_k > 0$ such that $T^{n_k} B \cap \bigcup_{i=0}^{k-1} T^i B = \emptyset$.   □

Proposition 1.2.6. *Let T be a measurable and nonsingular transformation defined on the finite measure space $(X, \mathscr{B}, m)$. If A is a compressible set of positive measure (i.e., $A \sim B$ for some subset $B \subset A$ and $m(A \smallsetminus B) > 0$) then there is a sequence of mutually disjoint sets $\{D_n : n \geq 1\}$ such that $D_i \approx D_j$ for $i \neq j$, $i, j = 1, 2, \ldots,$ and $m(D_1) > 0$.*

Proof. Since A is compressible there is a set $B \subset A$ with $m(A \smallsetminus B) > 0$ and subsets A_i, B_i and a sequence of integers $\{n_i : i \geq 1\}$, so that

$$A = \bigcup_{i=1}^{\infty} A_i \, (disj), \quad B = \bigcup_{i=1}^{\infty} B_i \, (disj), \quad \text{and} \quad B_i = T^{n_i} A_i \quad \text{for} \quad i \geq 1.$$

Let $C_1 = A \smallsetminus B$;   then $C_1 = \bigcup_{i=1}^{\infty} (A_i \cap C_1)(disj).$

Since

$$\bigcup_{i=1}^{\infty} T^{n_i} A_i \, (disj) = \bigcup_{i=1}^{\infty} B_i \, (disj) = B,$$

the set

$$C_2 = \bigcup_{i=1}^{\infty} T^{n_i} (A_i \cap C_1)(disj) \subset B = A \smallsetminus C_1.$$

We choose sets C_n for $n \geq 3$ inductively as follows. We assume $C_1, C_2, \ldots, C_{n-1}$ have been chosen with the property that

$$C_k = \bigcup_{i=1}^{\infty} T^{n_i} (A_i \cap C_{k-1}), \quad C_k \subset A \smallsetminus \bigcup_{j=1}^{k-1} C_j, \quad \text{and} \quad C_k \sim C_{k-1} \quad \text{for} \quad k \geq 2.$$

It follows that $C_n = \bigcup_{i=1}^{\infty} T^{n_i} (A_i \cap C_{n-1}) \, (disj) \subset A \smallsetminus \bigcup_{j=1}^{n-1} C_j$, and $C_n \sim C_{n-1}$. Thus we have a sequence of sets $\{C_n : n \geq 1\}$, $C_i \sim C_j$, and $C_i \cap C_j = \emptyset$ for $i, j \geq 1$ and $i \neq j$.

Next we let $\varepsilon > 0$ be such that $m(C_1) - \varepsilon > 0$. For each $k \geq 2$, since $C_1 \sim C_k$, it follows that

$$C_1 = \bigcup_{i=1}^{\infty} E_{i,k} \, (disj), \quad C_k = \bigcup_{i=1}^{\infty} F_{i,k} \, (disj), \quad \text{and} \quad T^{n(i,k)} E_{i,k} = F_{i,k} \quad \text{for} \quad i \geq 1.$$

Since we are in a finite measure space, for each $k \geq 2$ there is an $N_k > 0$ such that

$$m \left(\bigcup_{i=N_k}^{\infty} E_{i,k} \right) < \frac{\varepsilon}{2^{k-1}}.$$

We let $D_1 = C_1 \setminus \bigcup_{k=2}^{\infty} \bigcup_{i=N_k}^{\infty} E_{i,k}$. Then,

$$m(D_1) \geq m(C_1) - \sum_{k=2}^{\infty} \frac{\varepsilon}{2^{k-1}} = m(C_1) - \varepsilon > 0.$$

Next for $k \geq 2$ we define

$$D_k = \bigcup_{i=1}^{N_k-1} T^{n(i,k)} (E_{i,k} \cap D_1). \tag{1.8}$$

Equation (1.8) and the fact that $D_1 = \bigcup_{i=1}^{N_k-1} (E_{i,k} \cap D_1)$ imply $D_1 \approx D_k$ for $k \geq 2$. Moreover, (1.8) implies $D_k \subset C_k$ for $k \geq 2$, and therefore $D_i \cap D_j = \emptyset$ for $i, j \geq 1$ and $i \neq j$. □

Proposition 1.2.7. *Condition* (**D1**) *implies condition* (**F**).

Proof. Assume condition (**F**) is not true. Then there is a measurable function $f > 0$, a sequence of integers $\{n_i : i \geq 0\}$ and a set A with $m(A) > 0$ such that

$$\sum_{i=0}^{\infty} f(T^{n_i} x) < \infty \quad \text{for } x \in A. \tag{1.9}$$

Since $m(X) < \infty$, for each $\varepsilon > 0$ there is a $\delta > 0$ and a set B such that $m(B) < \varepsilon$, and $f(x) \geq \delta$ for $x \in X \setminus B$. Equation (1.9) implies that for $x \in A$, $T^{n_i} x \in X \setminus B$ for finitely many i only. In other words, for almost all $x \in A$ there is an integer $N = N(x) > 0$ such that $T^{n_i} x \in B$ for $i \geq N$.

For $k \geq 1$ let $A(k) = \{x \in A : N(x) = k\}$. Then $A = \bigcup_{k=1}^{\infty} A(k)$, and $m(A) > 0$ implies that for any fixed η, $0 < \eta < m(A)$, there is $k_1 > 0$ such that if $A_1 = A(k_1)$ then $m(A \setminus A_1) \leq \eta/2$. Therefore, for $x \in A_1$, $T^{n_i} x \in B$ for all $i \geq k$, or $m(T^{n_i} A_1) \leq \varepsilon$ for all $i \geq k_1$.

From the above discussion we conclude: for $\varepsilon > 0$ and $\eta > 0$ there is a subset $A_1 \subset A$ such that $\lim_{i \to \infty} m(T^{n_i} A_1) \leq \varepsilon$ and $m(A \setminus A_1) \leq \eta/2$.

We repeat the above argument, and using induction choose sets $A \supset A_1 \supset \cdots$ as follows: For $p \geq 1$ let $\varepsilon_p = \varepsilon/p$ and $\eta_p = \eta/2^p$. Assume we have chosen the sets $A \supset A_1 \supset \cdots \supset A_p$ for some $p \geq 1$ that satisfy

$$\lim_{i \to \infty} m(T^{n_i} A_p) \leq \varepsilon_p \text{ and } m(A_{p-1} \smallsetminus A_p) \leq \eta_p . \tag{1.10}$$

We use the same argument as above and obtain a set $A_{p+1} \subset A_p$ that satisfies (1.10) with p replaced by $p + 1$.

Finally, we let $A' = \bigcap_{p=1}^{\infty} A_p$. It follows that

$$\lim_{n \to \infty} m(T^n A') \leq \lim_{i \to \infty} m(T^{n_i} A') \leq \lim_{i \to \infty} m(T^{n_i} A_p) \leq \varepsilon_p \to 0 \text{ as } p \to \infty,$$

and

$$m(A') \geq m\left(A \smallsetminus \bigcup_{p=1}^{\infty} \left(A_p \smallsetminus A_{p+1} \right) \right) \geq m(A) - \sum_{p=1}^{\infty} \eta/2^p = m(A) - \eta > 0.$$

This is a contradiction to condition **(D1)**.   □

Let us introduce two more conditions on the transformation T:

(U) For any $\varepsilon > 0$ there is a $\delta > 0$ such that

$$m(A) < \delta \quad \Longrightarrow \quad m(T^n A) < \varepsilon \text{ for all } n \in \mathbb{Z}.$$

(W*) For any $\varepsilon > 0$ there is an integer $k > 0$ such that:
$m(A) > \varepsilon \Longrightarrow$ there are at most k mutually disjoint images of A by T.

(W*) is a stronger condition than condition **(W)**, and condition **(U)** will insure the countable additivity of the invariant measure.

Proposition 1.2.8. *Condition* **(D2)** *implies condition* **(W*)**.

Proof. We note that the set function $\overline{\sigma}(A) = \limsup_{n \to \infty} \frac{1}{n} \sum_{i=0}^{n-1} m(T^n A)$ is a subadditive monotonic set function. Using Proposition 1.2.2 we conclude that condition **(D2)** implies: for any $\varepsilon > 0$ there is a $\delta > 0$ such that

$$m(A) \geq \varepsilon \Longrightarrow \overline{\sigma}(A) = \limsup_{n \to \infty} \sigma_n(A) \geq \delta . \tag{1.11}$$

Let us assume condition **(W*)** does not hold. Then there is an $\varepsilon > 0$ such that for every positive integer $r > 0$ there is a set B and a collection $\{p_i : 0 \leq i < r\}$ of integers such that the sets $\{T^{p_i} B : 0 \leq i < r\}$ are mutually disjoint.

From (1.4) of Proposition 1.2.4 follows

$$r\overline{\sigma}(B) = \overline{\sigma}\left(\bigcup_{i=0}^{r-1} T^{p_i} B\right) \le m(X),$$

or

$$\overline{\sigma}(B) < \frac{m(X)}{r}.$$

Since $r > 0$ is arbitrary and $m(X) < \infty$, the above contradicts (1.11). □

Proposition 1.2.9. *Condition* (**W***) *implies condition* (**U**).

Proof. Assume condition (**U**) does not hold. Then there is an $\varepsilon > 0$, a sequence of sets $\{A_i : i \ge 1\}$, and a sequence of integers $\{n_i : i \ge 1\}$ such that $m(A_i) < \frac{1}{2^i}$ and $m(T^{n_i} A_i) > 2\varepsilon$ for all $i \ge 1$.

Let $B_k = \bigcup_{i=k}^{\infty} A_i$ for $k \ge 1$. Then, $B_1 \supset B_2 \supset \cdots$ with

$$\lim_{n\to\infty} m(B_k) = 0 \quad \text{and} \quad m(T^{n_k} B_k) > 2\varepsilon \quad \text{for all } k \ge 1. \tag{1.12}$$

For any set $A \in \mathcal{B}$ the set function $\underline{\sigma}(A) = \liminf_{n\to\infty} \frac{1}{n} \sum_{i=0}^{n-1} m(T^i A)$ is a superadditive and nonnegative set function defined on $\mathcal{B}$. Therefore, by Proposition 1.2.3, for any $\delta > 0$ there is an integer $n_0 > 0$ such that

$$\underline{\sigma}(B_{n_0} \smallsetminus B_n) < \delta \text{ for all } n \ge n_0. \tag{1.13}$$

In (1.12) choose any integer $k \ge 1$ and set $k_0 = k$; then $m(T^{n_{k_0}} B_{k_0}) > 2\varepsilon$. Since T is nonsingular, choose a positive integer $k_1 > k_0$ such that $m(T^{n_{k_0}} B_{k_1}) < \varepsilon$. Let $A = T^{n_{k_0}} (B_{k_0} \smallsetminus B_{k_1})$. Then

$$m(A) = m(T^{n_{k_0}} B_{k_0}) - m(T^{n_{k_0}} B_{k_1}) > 2\varepsilon - \varepsilon = \varepsilon,$$

and by (1.13) $\underline{\sigma}(A) = \underline{\sigma}(B_{k_0} \smallsetminus B_{k_1}) < \delta$.

Therefore, for some $\varepsilon > 0$ and any $\delta > 0$ there exists a set A satisfying:

$$m(A) \ge \varepsilon \quad \text{and} \quad \underline{\sigma}(A) = \liminf_{n\to\infty} \frac{1}{n} \sum_{i=0}^{n-1} m(T^i A) < \delta. \tag{1.14}$$

Next, for any $k > 0$ we construct a set B with $m(B) \ge \varepsilon/2$ such that the sets $T^{p_i} B$ for $i = 0, 1, 2, \ldots, k$ are mutually disjoint.

We let $k > 0$ be an arbitrary positive integer, and choose a $\delta > 0$ such that $\frac{k(k+1)}{2}\delta < \varepsilon/2$. From the above there is a set A that satisfies (1.14).

We recall $\underline{\sigma}(A) = \liminf_{n\to\infty} \sigma_n(A) = \liminf_{n\to\infty} \frac{1}{n} \sum_{i=0}^{n-1} m(T^i A)$.

We let $p_0 = 0$, and using (1.14) we choose an integer $p_1 > p_0$ such that $m(T^{p_1-p_0} A) < \delta$. Let $0 < i \leq k$, and assume that the integers $p_0, p_1, \ldots, p_{i-1}$ have been chosen. Next we choose a positive integer $p_i > p_{i-1}$ such that

$$m\left(\bigcup_{j=0}^{i-1} T^{p_i - p_j} A \right) < i\delta. \tag{1.15}$$

This is possible since by Proposition 1.2.4

$$\underline{\sigma}\left(\bigcup_{j=0}^{i-1} T^{-p_j} A \right) = \liminf_{n\to\infty} \sigma_n\left(\bigcup_{j=0}^{i-1} T^{-p_j} A \right) \leq \liminf_{n\to\infty} \sum_{j=0}^{i-1} \sigma_n\left(T^{-p_j} A \right)$$

$$\leq i\underline{\sigma}(A) < i\delta.$$

This way we obtain $k+1$ integers $0 = p_0 < p_1 < \cdots < p_k$ such that (1.15) holds for $i = 1, 2, \ldots, k$.

Let us put $A' = \bigcup_{i=1}^{k} \bigcup_{j=0}^{i-1} T^{p_i - p_j} A$. Then

$$m(A') \leq \sum_{i=1}^{k} m\left(\bigcup_{j=0}^{i-1} T^{p_i - p_j} A \right) < \sum_{i=1}^{k} i\delta = \frac{k(k+1)}{2} \delta < \varepsilon/2.$$

If we let $B = A \smallsetminus A'$ then $m(B) \geq \varepsilon/2$, and the sets $T^{p_i} B$ for $i = 1, 2, \ldots, k$ are mutually disjoint. In fact, for $i > j$ we have:
$B \subset A$ and $T^{p_i-p_j} B \cap B \subset T^{p_i-p_j} A \cap (A \smallsetminus A') = \emptyset$.   □

Now we prove the Finite Invariant Measure Theorem.

Proof (Theorem 1.2.1). The implications **(D1)** $\Rightarrow$ **(C)** $\Rightarrow$ **(D2)** are obvious.

(D2) $\Rightarrow$ **(H):** Assume condition **(H)** is not true. Then there are sets $A \supset B$ such that $A \sim B$ and $m(A \smallsetminus B) > 0$. From Proposition 1.2.6 follows that there is a sequence of mutually disjoint sets $\{D_n : n \geq 1\}$ such that $m(D_1) > 0$ and $D_1 \approx D_k$ for $k \geq 2$. Then for any integer $r > 0$ we have: $\overline{\sigma}(D_1 \cup D_2 \cup \cdots \cup D_r) = r\overline{\sigma}(D_1)$. This implies $r\overline{\sigma}(D_1) \leq m(X)$. Since this is true for any integer $r > 0$ and $m(X) < \infty$ we conclude $\overline{\sigma}(D_1) = 0$. This contradicts condition **(D2)**.

(H) $\Rightarrow$ **(W):** If condition **(W)** is not true then there is a ww set C and a sequence of integers $\{n_i : i \geq 1\}$ satisfying $T^{n_i} C \cap T^{n_j} C = \emptyset$ for $i \neq j$.

We let $A = \bigcup_{i=1}^{\infty} A_i$ and $B = \bigcup_{i=1}^{\infty} B_i$ where $A_i = T^{n_i} C$ and $B_i = T^{n_i+1} C$ for $i \geq 1$. Then $B_i = T^{n_i+1-n_i} A_i$. Therefore $B \sim A$, $B \subset A$, and $m(A \smallsetminus B) = m(T^{n_1} C) > 0$. This contradicts condition $(\mathbf{H})$.

$(\mathbf{W}) \Rightarrow (\mathbf{S})$: Suppose condition $(\mathbf{S})$ is not true. Then there is a measurable set A of positive measure so that for every integer n_k there is an integer n_{k+1} such that

$$m\left(T^{n_{k+1}} A \cap \bigcup_{i=-n_k}^{n_k} T^i A\right) = 0.$$

Let $N = \bigcup_{k=1}^{\infty} \left(T^{n_{k+1}} A \cap \bigcup_{i=-n_k}^{n_k} T^i A\right)$. Then $m(N) = 0$, and the set $B = A \smallsetminus N$ is a ww set under the sequence $\{n_k : k \geq 1\}$. This proves $(\mathbf{W}) \Rightarrow (\mathbf{S})$.

Proposition 1.2.5 proves the implication $(\mathbf{S}) \Rightarrow (\mathbf{D1})$ and completes the proof that conditions $(\mathbf{D1})$, $(\mathbf{C})$, $(\mathbf{D2})$, $(\mathbf{H})$, $(\mathbf{W})$, and $(\mathbf{S})$ are mutually equivalent.

Proposition 1.2.7 is the implication $(\mathbf{D1}) \Rightarrow (\mathbf{F})$. Finally, the equivalence of all the conditions in the theorem requires only the proof of the implication $(\mathbf{F}) \Rightarrow (\mathbf{W})$.

We assume condition $(\mathbf{W})$ is not satisfied. Then there is a sequence of integers $\{n_i : i \geq 0\}$ and a set C of positive measure such that $T^{n_i} C \cap T^{n_j} C = \emptyset$ for $i \neq j$. We let

$$f(x) = \begin{cases} 1 & \text{if } x \in X \smallsetminus \bigcup_{i=0}^{\infty} T^{n_i} C, \\ 1/2^i & \text{if } x \in T^{n_i} C \quad \text{for } i > 0. \end{cases}$$

Then for the above sequence $\{n_i\}$ and for $x \in C$ we have

$$\sum_{i=1}^{\infty} f(T^{n_i} x) = \sum_{i=1}^{\infty} 1/2^i < \infty.$$

This is a contradiction to condition $(\mathbf{F})$.

To obtain the finite T-invariant measure $\mu \sim m$ we note that Proposition 1.2.8 together with Proposition 1.2.9 show that condition $(\mathbf{D2})$ implies condition $(\mathbf{U})$. This says that for any $\varepsilon > 0$ there is $\delta > 0$ such that $m(A) < \delta$ implies $m(T^n A) < \varepsilon$ for all $n \in \mathbb{Z}$. In other words, if we write $\rho_n(A) = m(T^n A)$ then the sequence $\{\rho_n : n \in \mathbb{Z}\}$ is equi-uniformly absolutely continuous with respect to m. We note that for each $A \in \mathscr{B}$ the set $\{\rho_n(A) : n \in \mathbb{Z}\}$ is a bounded sequence of real numbers. Let us put

$$\mu(A) = \text{Lim} \, \rho_n(A),$$

where Lim denotes a Banach limit. It is clear that μ is invariant and finitely additive on $\mathscr{B}$ with $\mu(X) = 1$. From the equi-uniform absolute continuity of $\{\rho_n : n \in \mathbb{Z}\}$ follows that μ is countably additive and equivalent with m. In fact for any $\varepsilon > 0$ we choose a positive number $\delta > 0$ such that $m(A) < \delta$ implies $\rho_n(A) < \varepsilon$ for all $n \in \mathbb{Z}$. From this follows that μ is countably additive and uniformly absolutely continuous with respect to m.

The above proves the sufficiency of the conditions mentioned in the theorem. For the necessity of the conditions we note that the existence of a finite invariant measure $\mu \sim m$ implies the non-existence of ww sets.   □

Chapter 2
Transformations with No Finite Invariant Measure

In this chapter we consider properties of transformations defined on a sigma-finite measure space $(X, \mathcal{B}, m)$ that do not preserve a finite measure $\mu \sim m$.

2.1 Measurable Transformations

Definition 2.1.1. Let T be a measurable and nonsingular transformation defined on the σ-finite measure space $(X, \mathcal{B}, m)$.

- For any infinite set of integers $\{n_i : i \geq 1\}$ and any set $A \in \mathcal{B}$ let us consider the following sequence of sets related to the set A and the sequence $\{n_i\}$:

Let $n_0 = 0$, $A_0 = A$, $A_1 = TA \smallsetminus \bigcup_{r=0}^{\infty} T^{n_r} (T^{-n_0} A_0)$,

and $A_2 = T^2 A \smallsetminus \bigcup_{r=0}^{\infty} T^{n_r} (T^{-n_0} A_0 \cup T^{-n_1} A_1)$.

We continue by induction. For $p > 2$ we let

$$A_p = T^p A \smallsetminus \bigcup_{r=0}^{\infty} T^{n_r} \left(T^{-n_0} A_0 \cup T^{-n_1} A_1 \cup \cdots \cup T^{-n_{p-1}} A_{p-1} \right). \qquad (2.1)$$

Let us call the set

$$W = \bigcup_{p=0}^{\infty} T^{-n_p} A_p \qquad (2.2)$$

the *derived set* from the sequence $\{n_i : i \geq 1\}$ and the set $A \in \mathcal{B}$.

© Springer Japan 2014
S. Eigen et al., *Weakly Wandering Sequences in Ergodic Theory*,
Springer Monographs in Mathematics, DOI 10.1007/978-4-431-55108-9_2

We recall Definition 1.1.1. An infinite set of integers $\{n_i : i \geq 0\}$ is a *ww* sequence for T if there is a set W of positive measure such that $T^{n_i} W \cap T^{n_j} W = \emptyset$ for $i, j \geq 0$ and $i \neq j$.

- An infinite set of integers $\{n_i : i \geq 0\}$ is called an *exhaustive weakly wandering (eww)* sequence for T if there is a set W such that $X = \bigcup_{i=0}^{\infty} T^{n_i} W$ (*disj*). Often we will say W is an *eww* set (for T) (with the sequence $\{n_i\}$), or $\{n_i\}$ is an *eww* sequence (for T) (with the set W).
- An infinite set of integers $\{n_i : i \geq 0\}$ is a *strongly weakly wandering (sww)* sequence for T (with the set A) if $m(A) > 0$ and for $i, j, k, l \geq 0$, $i > j$ we have: $T^{n_i - n_k + k} A \cap T^{n_j - n_l + l} A = \emptyset$ whenever one of the indices $\{i, j, k, l\}$ is larger than all the others, or $i = l > \max\{j, k\}$.

The following simple proposition follows from the above definitions.

Proposition 2.1.2. *Let $\{n_i : i \geq 0\}$ (where $n_0 = 0$) be an sww sequence for the measurable and nonsingular transformation T with the set $A \in \mathcal{B}$. Then the derived set W from the sequence $\{n_i\}$ and the set A satisfies:*

$$\bigcup_{r=0}^{\infty} T^{n_r} W (disj) \supset \bigcup_{p=1}^{\infty} T^p A .$$

Proof. From (2.1) and (2.2) in the definition of the derived set follows that for all $p > 0$,

$$\bigcup_{r=0}^{\infty} T^{n_r} W \supset A_p \cup \bigcup_{r=0}^{\infty} T^{n_r} \left(T^{-n_0} A_0 \cup T^{-n_1} A_1 \cup \cdots \cup T^{-n_{p-1}} A_{p-1} \right)$$

$$\supset T^p A .$$

This implies $\displaystyle\bigcup_{r=0}^{\infty} T^{n_r} W \supset \bigcup_{p=0}^{\infty} T^p A .$

It remains to show: $T^{n_i} W \cap T^{n_j} W = \emptyset$ for $i, j \geq 0$ and $i > j$. For this it is sufficient to show that

$$T^{n_i - n_k} A_k \cap T^{n_j - n_l} A_l = \emptyset \quad \text{for} \quad i, j, k, l \geq 0, \text{ and } i > j. \tag{2.3}$$

It is clear from (2.1) that for any integer $r > 0$

$$A_p \cap T^{n_r - n_s} A_s = \emptyset \quad \text{if } p > s. \tag{2.4}$$

If $i = k > \max\{j, l\}$ then (2.3) follows from (2.4). In all the other cases we note that $A_k \subset T^k A$ for $k \geq 0$, and (2.3) follows from the properties defining the *sww* sequence $\{n_i\}$.   □

Proposition 2.1.3. *Let T be a measurable and nonsingular transformation defined on the σ-finite measure space $(X, \mathcal{B}, m)$. If T does not preserve a finite measure μ equivalent to m, then there exists an infinite sequence $\{n_i : i \geq 0\}$ of integers with the property that every infinite subsequence of the sequence $\{n_i : i \geq 0\}$ is an sww sequence for T.*

For a set $A \in \mathcal{B}$ and for an integer $p \geq 0$ let $A^p = \bigcup\limits_{j=-p}^{p} T^j A$.

We first prove the following lemma.

Lemma 2.1.4. *Let T be a measurable and nonsingular transformation defined on the σ-finite measure space $(X, \mathcal{B}, m)$. Suppose there exists a set A of positive measure such that*

$$\liminf_{n \to \infty} \left[m(T^n A^p \cap A) + m(T^{2n} A^p \cap A) \right] = 0 \ \text{ for all } \ p \geq 0. \qquad (2.5)$$

Then there exists an infinite sequence $\{n_i : i \geq 0\}$ of integers with the property that every infinite subsequence of the sequence $\{n_i : i \geq 0\}$ is an sww sequence for T.

Proof. Let $0 < \varepsilon < m(A)$, and for $i \geq 1$ let $\varepsilon_i = \varepsilon/2^i$.

We let $n_0 = 0$, $p_1 = 1$ and consider the set $A^{p_1} = \bigcup_{s=-p_1}^{p_1} T^s A$. Using (2.5) we choose an integer $n_1 > n_0$ such that $m\left(T^{n_1} A^{p_1} \cap A\right) + m\left(T^{2n_1} A^{p_1} \cap A\right) < \varepsilon_1$. Next we let $p_2 = 2n_1 + p_1 + 2$, and consider $A^{p_2} = \bigcup_{s=-p_2}^{p_2} T^s A$. We use (2.5) and choose an integer $n_2 > n_1$ such that $m\left(T^{n_2} A^{p_2} \cap A\right) + m\left(T^{2n_2} A^{p_2} \cap A\right) < \varepsilon_2$.

We continue by induction. Assume the integers $n_1, n_2, \ldots, n_{k-1}$ and $p_1, p_2, \ldots, p_{k-1}$ have been chosen. We let $p_k = 2n_{k-1} + p_{k-1} + k$ and consider the set $A^{p_k} = \bigcup_{s=-p_k}^{p_k} T^s A$. We use (2.5) to choose an integer $n_k > n_{k-1}$ so that

$$m\left(T^{n_k} A^{p_k} \cap A\right) + m\left(T^{2n_k} A^{p_k} \cap A\right) < \varepsilon_k.$$

Finally, we let

$$A_0 = A \smallsetminus \bigcup_{s=1}^{\infty} \left(T^{n_s} A^{p_s} \cap A\right) \cup \left(T^{2n_s} A^{p_s} \cap A\right).$$

We note that

$$T^{n_s} A^{p_s} \cap A_0 = \emptyset \ \text{ and } \ T^{2n_s} A^{p_s} \cap A_0 = \emptyset \ \text{ for all } s \geq 1. \qquad (2.6)$$

We have

$$m(A_0) \geq m(A) - m\left[\bigcup_{s=1}^{\infty} (T^{n_s} A^{p_s} \cap A) \cup (T^{2n_s} A^{p_s} \cap A)\right]$$

$$\geq m(A) - \sum_{s=1}^{\infty} \left[m\left(T^{n_s} A^{p_s} \cap A\right) + m\left(T^{2n_s} A^{p_s} \cap A\right)\right] > m(A) - \varepsilon > 0.$$

We examine when the following is true:

$$\text{for} \quad i, j, k, l \geq 0, \ 0 \leq k' \leq k, \ 0 \leq l' \leq l \ \text{and} \ i > j$$

$$T^{n_i - n_k + k'} A_0 \cap T^{n_j - n_l + l'} A_0 = \emptyset. \tag{2.7}$$

If one of the indices, say k, is larger than all the others, then

$$T^{n_k + k' - n_i + n_j - n_l - l'} A_0 \subset T^{n_k} A^{p_k},$$

and (2.6) implies (2.7).

If $i = l > \max\{j, k\}$, then $T^{2n_i - n_k + k' + n_j - l'} A_0 \subset T^{2n_i} A^{p_i}$ and again (2.6) implies (2.7).

Next we show that any infinite subsequence of the sequence $\{n_i : i \geq 1\}$ is an sww sequence for the transformation T with the same set A_0.

Let us consider an infinite subsequence of the sequence $\{n_i : i \geq 0\}$ after eliminating some of its members. Let $n_0 = 0$, and after re-indexing the new sequence in the order obtained let us denote it by the same symbols $\{n_i : i \geq 0\}$. From (2.7) follows that for this sequence $\{n_i\}$ the following holds:

for $\quad i, j, k, l \geq 0, \ i > j \quad T^{n_i - n_k + k} A_0 \cap T^{n_j - n_l + l} A_0 = \emptyset \quad$ whenever one of the indices $\{i, j, k, l\}$ is larger than all the others, or $i = l > \max\{j, k\}$. $\qquad\square$

Proof (Proposition 2.1.3).

Since T does not preserve a finite equivalent measure, Theorem 1.2.1 implies that there exists a set A of positive measure with $\lim_{n \to \infty} \frac{1}{n} \sum_{i=0}^{n-1} m(T^i A) = 0$. We assume $0 < m(A) < \infty$. Then it is clear that $\lim_{n \to \infty} \frac{1}{n} \sum_{i=0}^{n-1} m(T^i A \cap A) = 0$.

We note that for $p \geq 0$

$$\lim_{n \to \infty} \frac{1}{n} \sum_{i=0}^{n-1} m(T^i A^p \cap A) = \lim_{n \to \infty} \frac{1}{n} \sum_{i=0}^{n-1} m\left[T^i \left(\bigcup_{j=-p}^{p} T^j A\right) \cap A\right]$$

$$\leq \sum_{j=-p}^{p} \lim_{n \to \infty} \frac{1}{n} \sum_{i=0}^{n-1} m\left(T^{i+j} A \cap A\right)$$

$$= (2p + 1) \lim_{n \to \infty} \frac{1}{n} \sum_{i=0}^{n-1} m\left(T^i A \cap A\right) = 0.$$

The above together with the inequality

$$\frac{1}{n} \sum_{i=0}^{n-1} \left[m(T^i A^p \cap A) + m(T^{2i} A^p \cap A) \right] \le \frac{1}{2n} \sum_{i=0}^{2n-1} 4m(T^i A^p \cap A)$$

imply

$$\lim_{n \to \infty} \frac{1}{n} \sum_{i=0}^{n-1} \left[m(T^i A^p \cap A) + m(T^{2i} A^p \cap A) \right] = 0.$$

This in turn implies:

$$\liminf_{i \to \infty} \left[m(T^n A^p \cap A) + m(T^{2n} A^p \cap A) \right] = 0 \quad \text{for all} \quad p \ge 0.$$

Lemma 2.1.4 then completes the proof. $\square$

Some transformations that do not preserve a finite measure $\mu \sim m$ preserve a σ-finite infinite measure. Therefore, we shall assume from here on that the measure space $(X, \mathscr{B}, m)$ we consider is σ-finite and not necessarily finite.

2.2 Ergodic Transformations

At this point we impose a new restriction on the measurable and nonsingular transformations. We say that the transformation T is *ergodic* if the only T-invariant sets are trivial, that is

$$TA = A \implies m(A) = 0 \quad \text{or} \quad m(X \smallsetminus A) = 0.$$

In the following proposition we list a few properties of ergodic transformations.

Proposition 2.2.1 (Ergodic). *Let T be an ergodic transformation defined on the σ-finite measure space $(X, \mathscr{B}, m)$. Then T satisfies the following properties:*

(1E) $T^i A \cap T^j A = \emptyset$ *for $i \ne j$* $\implies m(A) = 0$ *(i.e. T does not accept wandering sets).*

(2E) $TA \subset A \implies m(A \smallsetminus TA) = 0.$

(3E) $m(A) > 0 \implies$ *for any integer $k \ge 0$ we have $\bigcup_{n=k}^{\infty} T^n A = X$.*

(4E) f *a measurable function,* $f(Tx) = f(x)$ *a.e.* $\implies f(x) \equiv c$ *for some constant c.*

(5E) *If $m \sim \mu$, and both m and μ are invariant measures for T, then $m = c\mu$ for some constant $c \ne 0$.*

Proof. Assume property (1E) is not true, and let A be a wandering set of positive measure. Let $B \subset A$ such that $0 < m(B) < m(A)$. Then the set

$B^* = \bigcup_{n=-\infty}^{\infty} T^n B$ has the property that $TB^* = B^*$ and $m(B^*) > 0$ with $m(X \setminus B^*) \geq m(A \setminus B^*) > 0$. This is a contradiction to T being ergodic.

Assume property (2E) is not true. Let A be a set with $TA \subset A$ and $m(A \setminus TA) > 0$. Then the set $B = A \setminus TA$ is a wandering set. This is a contradiction to property (1E).

For any integer $k \geq 0$ and any set A of positive measure, let $A^* = \bigcup_{n=k}^{\infty} T^n A$. Then property (2E) implies $TA^* = A^*$. T ergodic implies that $A^* = X$, and this proves property (3E).

Assume property (4E) is not true, and let f be a measurable function satisfying $f(Tx) = f(x)$ a.e. Let c be a constant such that the sets $A = \{x : f(x) > c\}$ and $B = \{x : f(x) < c\}$ both have positive measure. It follows that $TA = A$, $TB = B$ and $A \cap B = \emptyset$. This is a contradiction to T being ergodic.

Suppose T preserves two measures m and μ, $m \sim \mu$. Let $f(x)$ be the Radon–Nikodym derivative of m with respect to μ. Then for every $A \in \mathscr{B}$,

$$\int_A f(Tx)d\mu(x) = \int_A f(Tx)d\mu(Tx) = \int_{TA} f(x)d\mu(x)$$

$$= m(TA) = m(A) = \int_A f(x)d\mu(x).$$

This implies $f(Tx) = f(x)$, and property (4E) says $f(x) \equiv c$ for some constant $c \neq 0$. This proves property (5E).   □

Ergodic transformations are the basic building blocks for nonsingular transformations, and in the literature there are many examples of ergodic transformations, some preserving a finite measure equivalent to m, others preserving a σ-finite infinite measure equivalent to m, and some preserving no σ-finite measure equivalent to m. From Proposition 2.2.1 property (1E) follows that ergodic transformations are recurrent. Property (4E) of Proposition 2.2.1 implies that if an ergodic transformation preserves an infinite measure m then it does not preserve a finite measure $\mu \sim m$; in particular, none of the conditions (1R)–(6R) of the Recurrence Theorem 1.1.3 is sufficient for the existence of a finite invariant measure $\mu \sim m$ (see also [29]). The Finite Invariant Measure Theorem 1.2.1 then implies: if an ergodic transformation T preserves an infinite measure then it necessarily accepts ww sequences. This is in sharp contrast to the misconception that an ergodic transformation in general has some sort of mixing character. It turns out that ergodic transformations that do not preserve a finite measure $\mu \sim m$ admit even more interesting sequences than the weakly wandering ones, as we exhibit below.

We note that for a given transformation T it is easier to verify the "disjointness" condition in the definition of sww sequences than it is to verify the "exhaustive" condition of eww sequences. The next proposition shows that for an ergodic transformation T all the sww sequences are eww.

Proposition 2.2.2. *Let T be an ergodic transformation. Then every sww sequence for T is an eww sequence for T.*

Proof. The proof follows from Proposition 2.1.2 and property (**3E**) of Proposition 2.2.1.   □

The next theorem exhibits the existence of a special kind of *eww* sequence for an ergodic transformation that does not accept a finite invariant measure $\mu \sim m$. The following theorem generalizes a result of Jones and Krengel [42].

Theorem 2.2.3. *Let T be an ergodic transformation defined on the σ-finite measure space $(X, \mathscr{B}, m)$. If T does not preserve a finite measure $\mu \sim m$, then it possesses an eww sequence $\{n_i\}$ with the property that every infinite subsequence $\{n_i'\}$ of the sequence $\{n_i\}$ is again an eww sequence for the transformation T.*

Proof. From Proposition 2.1.3 follows that there exists an *sww* sequence $\{n_i\}$ for the transformation T. Moreover, every subsequence $\{n_i'\}$ of the *sww* sequence $\{n_i\}$ is again an *sww* sequence. Proposition 2.2.2 then concludes the proof.   □

In general it is not clear if or when a *ww* sequence contains an *eww* subsequence. We complete this chapter with the following proposition, which provides an interesting condition for a *ww* sequence to contain an *eww* subsequence.

Proposition 2.2.4. *If both $\{n_i\}$ and $\{2n_i\}$ are ww sequences for the ergodic transformation T, then the sequence $\{n_i\}$ contains an eww subsequence $\{n_i'\}$ with the additional property that every infinite subsequence of the sequence $\{n_i'\}$ is again an eww sequence for T.*

Proof. Let $\{n_i\}$ be a *ww* sequence for the transformation T with the set C. We may assume that $0 < m(C) < \infty$ and consider the set $C^p = \bigcup_{s=-p}^{p} T^s C$ for $p \geq 0$. For any integer $s \in \mathbb{Z}$ the sets $T^{n_i}(T^s C)$ are mutually disjoint for $i \geq 0$. Then

$$m(C) \geq m\left(\bigcup_{i=1}^{\infty} T^{n_i}(T^s C) \cap C\right) = \sum_{i=1}^{\infty} m\left(T^{n_i}(T^s C) \cap C\right)$$

implies $\lim_{i \to \infty} m\left(T^{n_i}(T^s C) \cap C\right) = 0$. Then

$$\lim_{i \to \infty} m\left(T^{n_i} C^p \cap C\right) = \lim_{i \to \infty} m\left(T^{n_i} \bigcup_{s=-p}^{p} T^s C \cap C\right)$$

$$\leq \sum_{s=-p}^{p} \lim_{i \to \infty} m\left(T^{n_i}(T^s C \cap C)\right) = 0.$$

Similarly, since $\{2n_i\}$ is a *ww* sequence for T, there is a set D of positive measure such that if $D^p = \bigcup_{s=-p}^{p} T^s D$ for $p \geq 0$ then $\lim_{i \to \infty} m\left(T^{2n_i} D^p \cap D\right) = 0$.

T ergodic implies: for some integer $k > 0$ the set $A = T^k C \cap D$ has positive measure, and both sequences $\{n_i\}$ and $\{2n_i\}$ are ww sequences for the transformation T with the same set A. From the above discussion we conclude that for $p \geq 0$ the sets $A^p = \bigcup_{s=-p}^{p} T^s A$ satisfy

$$\lim_{i \to \infty} [m(T^{n_i} A^p \cap A) + m(T^{2n_i} A^p \cap A)] = \lim_{i \to \infty} m(T^{n_i} A^p \cap A) + \lim_{i \to \infty} m(T^{2n_i} A^p \cap A) = 0.$$

Lemma 2.1.4 and Proposition 2.2.2 then complete the proof. $\square$

Chapter 3
Infinite Ergodic Transformations

In the previous chapters we saw that recurrent transformations do not accept wandering sets. An important subset of the recurrent transformations are the ergodic ones that do not have a finite invariant and equivalent measure. These transformations also do not accept wandering sets, yet they must necessarily accept *ww* and *eww* sets. For infinite ergodic transformations the existence of *ww* sets is a significant property and reflects the subtle features of these transformations. In this chapter we discuss the strong bond that exists between infinite ergodic transformations and *ww* or *eww* sequences.

3.1 General Properties of Infinite Ergodic Transformations

Let T be an ergodic transformation defined on the σ-finite measure space $(X, \mathscr{B}, m)$. Let us recall the definition: T is *ergodic* if $TA = A$ implies either $m(A) = 0$ or $m(X \smallsetminus A) = 0$; in other words, the only invariant sets under T are sets of measure zero or their complements. An equivalent description of an ergodic transformation T is the following:

$$m(A) > 0, \ m(B) > 0 \implies \exists \, n > 0 \quad \text{such that} \quad m(T^n A \cap B) > 0 . \qquad (3.1)$$

This says that for any two sets A and B of positive measure there exists an integer $n > 0$ such that the nth iterate of the images under T of one set intersects the other in a set of positive measure.

We say that an ergodic transformation T is a *finite ergodic* transformation if it preserves a finite measure $\mu \sim m$, and it is an *infinite ergodic* transformation if it preserves an infinite measure $\mu \sim m$.

We saw in Proposition 2.2.1 that the set of infinite ergodic transformations and the set of finite ergodic transformations are mutually exclusive. In this chapter we show that infinite ergodic transformations differ in more fundamental ways from the finite ergodic ones.

© Springer Japan 2014
S. Eigen et al., *Weakly Wandering Sequences in Ergodic Theory*,
Springer Monographs in Mathematics, DOI 10.1007/978-4-431-55108-9_3

Let T be a finite ergodic transformation defined on $(X, \mathscr{B}, m)$ with $m(X) = 1$. The following well-known property is an immediate consequence of Birkhoff's Individual Ergodic Theorem:

$$A, B \in \mathscr{B} \implies \lim_{n \to \infty} \frac{1}{n} \sum_{i=0}^{n-1} m(T^i A \cap B) = m(A)m(B). \qquad (3.2)$$

For finite ergodic transformations property (3.2) is a significant property, and it is sharper than property (3.1). It implies that, on the average, finite ergodic transformations are "mixing." This was an exciting result when it was first noticed, and at times it led to a feeling among some mathematicians (mixing gin and vermouth) that ergodic transformations possess some "mixing" feature in general. Consequently, property (3.2) was strengthened to:

$$A, B \in \mathscr{B} \implies \lim_{n \to \infty} m(T^n A \cap B) = m(A)m(B). \qquad (3.3)$$

Transformations with the above property were called "strongly mixing" or at times simply "mixing" transformations. Some important examples and many significant results followed. However, this exciting feeling of "mixing" for ergodic transformations made it quite difficult to suspect the existence of ww sequences for infinite ergodic transformations in general; see [38], Sect. 16.

3.2   Weakly Wandering Sequences

In this section we shall concentrate on infinite ergodic transformations. For such transformations the following property analogous to the important property (3.2) for finite ergodic transformations is well known, and it too is an immediate consequence of Birkhoff's Individual Ergodic Theorem:

$$m(A) < \infty, \ m(B) < \infty \implies \lim_{n \to \infty} \frac{1}{n} \sum_{i=0}^{n-1} m(T^i A \cap B) = 0. \qquad (3.4)$$

Unlike property (3.2) for finite ergodic transformations, property (3.4) for infinite ergodic transformations has not received proper attention. However, it is a significant property, and it is crucial in showing many of the interesting features that infinite ergodic transformations possess. Because of the significance of property (3.4) for infinite ergodic transformations we present a direct proof of it. We prove it using direct and elementary arguments avoiding the full power of Birkhoff's Individual Ergodic Theorem.

Theorem 3.2.1. *Let T be an infinite ergodic transformation defined on the infinite measure space $(X, \mathscr{B}, m)$. Then property (3.4) holds. Namely,*

$$m(A) < \infty, \ m(B) < \infty \implies \lim_{n\to\infty} \frac{1}{n} \sum_{i=0}^{n-1} m(T^i A \cap B) = 0.$$

We first prove the following two lemmas.

Lemma 3.2.2. *Let T be an infinite ergodic transformation defined on the infinite measure space $(X, \mathcal{B}, m)$. Then any two sets $C, D \in \mathcal{B}$ with the same finite measure are countably equivalent, that is,*

$$m(C) = m(D) < \infty \implies C \sim D.$$

Proof. Let $C_0 = D_0 = C \cap D$ and $n_0 = 0$.

In the case where $m(C \smallsetminus C_0) = m(D \smallsetminus D_0) = 0$, we let $C_i = D_i = \emptyset$ and $n_i = 0$ for all $i > 0$. Otherwise we let $D_1 = T^{p_1}(C \smallsetminus C_0) \cap (D \smallsetminus D_0)$, where p_1 is the smallest positive integer such that $m[T^{p_1}(C \smallsetminus C_0) \cap (D \smallsetminus D_0)] > 0$, and we let $C_1 = T^{-p_1} D_1$.

We continue by induction. For an integer $k > 0$, suppose we have chosen the subsets $C_0, C_1, C_2, \ldots, C_k$ of C, the subsets $D_0, D_1, D_2, \ldots, D_k$ of D and the integers $p_0, p_1, p_2, \ldots, p_k$ that satisfy the following:

for each $0 \le i \le k$, $C_i \subset C \smallsetminus \bigcup_{j=0}^{i-1} C_j$, $D \smallsetminus \bigcup_{j=0}^{i-1} D_j$, and $C_i = T^{-p_i} D_i$ where p_i is the smallest positive integer such that $m[T^{p_i}(C \smallsetminus \bigcup_{j=0}^{i-1} C_j) \cap (D \smallsetminus \bigcup_{j=0}^{i-1} D_j)] > 0$.

In the case where $m(C \smallsetminus \bigcup_{j=0}^{k} C_j) = m(D \smallsetminus \bigcup_{j=0}^{k} D_j) = 0$, we let $C_i = D_i = \emptyset$ for all $i > k$. Otherwise we let $D_{k+1} = T^{p_{k+1}}(C \smallsetminus \bigcup_{j=0}^{k} C_j) \cap (D \smallsetminus \bigcup_{j=0}^{k} D_j)$, where p_{k+1} is the smallest positive integer such that $m[T^{p_{k+1}}(C \smallsetminus \bigcup_{j=0}^{k} C_j) \cap (D \smallsetminus \bigcup_{j=0}^{k} D_j)] > 0$, and we let $C_{k+1} = T^{-p_{k+1}} D_{k+1}$.

Since $m(C) = m(D) < \infty$, we finally conclude that $C = \bigcup_{i=0}^{\infty} C_i$ *(disj)* and $D = \bigcup_{i=0}^{\infty} D_i$ *(disj)*, where for $i = 0, 1, 2, \ldots$ we have $D_i = T^{p_i} C_i$ for $p_i \in \mathbb{Z}$. □

Let B be a set of finite measure, and for any set $E \in \mathcal{B}$ consider the set functions

$$\sigma_{B,n}(E) = \frac{1}{n} \sum_{i=0}^{n-1} m(T^i B \cap E) \quad \text{and} \quad \overline{\sigma}_B(E) = \limsup_{n\to\infty} \sigma_{B,n}(E).$$

The next lemma is similar to Proposition 1.2.4 and exhibits the additive feature of the subadditive set function $\overline{\sigma}_B$.

Lemma 3.2.3. *Let T be an ergodic measure-preserving transformation defined on the infinite measure space $(X, \mathcal{B}, m)$. Let $A_0, A_1, \ldots, A_{r-1}$ be a finite collection of sets which are countably equivalent with each other, and let $B \in \mathcal{B}$ be a set of finite measure. Then*

$$\limsup_{n\to\infty} \sum_{i=0}^{r-1} \sigma_{B,n}(A_i) = r\overline{\sigma}_B(A_0). \tag{3.5}$$

In particular, if the sets $A_0, A_1, \ldots, A_{r-1}$ are mutually disjoint then

$$\overline{\sigma}_B\left(\bigcup_{p=0}^{r-1} A_p\right) = r\overline{\sigma}_B(A_0). \tag{3.6}$$

Proof. We observe that for any set $C \in \mathscr{B}$ and integers $p \in \mathbb{Z}$ and $n \in \mathbb{N}$, we have

$$\left|\sigma_{B,n}(C) - \sigma_{B,n}(T^p C)\right| = \left|\frac{1}{n}\sum_{i=0}^{n-1} m\left(T^i C \cap B\right) - \frac{1}{n}\sum_{i=0}^{n-1} m\left(T^{i+p} C \cap B\right)\right|$$

$$= \frac{1}{n}\left|\sum_{i=0}^{n-1} m\left(T^i C \cap B\right) - \sum_{i=p}^{p+n-1} m\left(T^i C \cap B\right)\right|$$

$$\leq \frac{2|p|}{n} m(B). \tag{3.7}$$

Let $C \sim D$; then

$$C = \bigcup_{j=1}^{\infty} C_j \,(disj), \ D = \bigcup_{j=1}^{\infty} D_j \,(disj) \text{ and } T^{p_j} C_j = D_j \text{ for } j \geq 1.$$

Since $m(C) < \infty$ and $m(D) < \infty$, for any $\varepsilon > 0$ there is an integer $N > 0$ such that

$$m\left(\bigcup_{j=N}^{\infty} C_j\right) < \varepsilon \text{ and } m\left(\bigcup_{j=N}^{\infty} D_j\right) < \varepsilon.$$

Let $C' = \bigcup_{j=0}^{N-1} C_j, \quad C'' = \bigcup_{j=N}^{\infty} C_j, \quad \text{and} \quad D' = \bigcup_{j=0}^{N-1} D_j, \quad D'' = \bigcup_{j=N}^{\infty} D_j.$

From (3.7) follows that for any sequence of integers n_k increasing to ∞,

$$\left|\sigma_{B,n_k}(C) - \sigma_{B,n_k}(D)\right| = \left|\sigma_{B,n_k}(C' \cup C'') - \sigma_{B,n_k}(D' \cup D'')\right|$$

$$\leq \left|\sum_{j=0}^{N-1}\left[\sigma_{B,n_k}(C_j) - \sigma_{B,n_k}(T^{p_j} C_j)\right]\right| + \left|\sigma_{B,n_k}(C'')\right|$$

$$+ \left|\sigma_{B,n_k}(D'')\right|$$

$$\leq \sum_{j=0}^{N-1} \frac{2|p_j|}{n_k} m(B) + 2\varepsilon,$$

and this implies $\overline{\sigma}_B(C) = \overline{\sigma}_B(D)$. Therefore we have

$$\left| \sum_{i=0}^{r-1} \sigma_{B,n_k}(A_i) - r\sigma_{B,n_k}(A_0) \right| \leq \sum_{i=0}^{r-1} \left| \sigma_{B,n_k}(A_i) - \sigma_{B,n_k}(A_0) \right| \longrightarrow 0 \qquad (3.8)$$

as $n_k \longrightarrow \infty$. Then (3.5) follows from (3.8), and (3.6) is a consequence of (3.5).   □

Proof (Theorem 3.2.1). Suppose $0 < m(A) < \infty$ and $0 < m(B) < \infty$.

Let $A_0 = A$, and let $r > 0$ be a positive integer. Since $(X, \mathcal{B}, m)$ is a non-atomic infinite measure space and $0 < m(A_0) < \infty$, it is possible to find r mutually disjoint sets, $A_0, A_1, \ldots, A_{r-1}$ such that $m(A_0) = m(A_i)$ for $1 \leq i \leq r-1$.

Using Lemma 3.2.2 we get $A_i \sim A_j$ for $i, j = 0, 1, 2, \ldots, r-1$. Then from Lemma 3.2.3 follows:

$$\overline{\sigma}_B\left(\bigcup_{0 \leq i < r} A_i \right) = r\overline{\sigma}_B(A_0), \quad \text{or} \quad \overline{\sigma}_B(A) \leq \tfrac{1}{r}\overline{\sigma}_B(X) = \tfrac{1}{r}m(B).$$

Since $m(B) < \infty$, this says $\overline{\sigma}_B(A) = 0$ or $\lim_{n\to\infty} \tfrac{1}{n} \sum_{i=0}^{n-1} m(T^i A \cap B) = 0$.
                                                                                       □

It is clear that property (3.4) implies:

$$m(A) < \infty, \; m(B) < \infty \implies \liminf_{n\to\infty} m(T^n A \cap B) = 0, \; \text{or equivalently:}$$

For any two sets A and B of finite measure, and any $\varepsilon > 0$ there exist arbitrarily large integers $n > 0$ such that $m(T^n A \cap B) < \varepsilon$.

Loosely speaking, since the whole space is infinite, the above property says that iterates of sets of finite measure under powers of T become almost disjoint. The above property of an infinite ergodic transformation is in a sense dual to the following basic property of a general ergodic transformation:

For any two sets A and B of positive measure there exist arbitrarily large integers $n > 0$ such that $m(T^n A \cap B) > 0$.

Another significant property that an infinite ergodic transformation T possesses is the fact that it admits *ww* sequences. As we saw earlier this follows from the Finite Invariant Measure Theorem 1.2.1. For infinite ergodic transformations this fact was not suspected until *ww* sets were introduced and shown to exist in [32] for every infinite ergodic transformation.

For an infinite ergodic transformation the two properties mentioned above, namely property (3.4) and the existence of *ww* sequences are basic and important properties. They give a general description of the geometric nature of infinite ergodic transformations and are useful in exhibiting further properties of such transformations. In the next proposition, using direct and elementary arguments, we show their equivalence (see [28]).

Proposition 3.2.4. *Let T be an ergodic measure-preserving transformation defined on the infinite measure space* $(X, \mathscr{B}, m)$. *Then property (3.4), namely*

$$m(A) < \infty, \ \ m(B) < \infty \ \ \Longrightarrow \ \ \lim_{n \to \infty} \frac{1}{n} \sum_{i=0}^{n-1} m(T^i A \cap B) = 0$$

holds if and only if T possesses ww sequences.

Lemma 3.2.5. *Let T be an ergodic measure-preserving transformation defined on the infinite measure space* $(X, \mathscr{B}, m)$, *and let* $\{v_i\}$ *be an infinite sequence of integers satisfying:*

$$m(A) < \infty \ \ \Longrightarrow \ \ \liminf_{i \to \infty} m(T^{v_i} A \cap A) = 0. \tag{3.9}$$

Then the sequence $\{v_i\}$ *contains a ww subsequence* $\{w_i\}$.

Proof. Let A and B be two sets of finite measure. Then

$$\liminf_{i \to \infty} m[T^{v_i} (A \cup B) \cap (A \cup B)] \geq \liminf_{i \to \infty} m(T^{v_i} B \cap A).$$

Applying (3.9) to the set $A \cup B$ we conclude from the above:

$$m(A) < \infty, \ \ m(B) < \infty \ \ \Longrightarrow \ \ \liminf_{i \to \infty} m(T^{v_i} B \cap A) = 0. \tag{3.10}$$

Let $0 < m(A) < \infty$. For any $\varepsilon > 0$ with $\varepsilon < m(A)$ let $\varepsilon_i = \varepsilon / 2^i$ for $i = 1, 2, \ldots$. Without any loss of generality we assume $\{v_i\}$ consists of only positive integers.

Let $w_0 = 0$. From (3.10) follows that there exists an integer $w_1 > 0$ such that $m(T^{w_1} A \cap A) < \varepsilon_1$.

We have $m(A \cup T^{-w_1} A) < \infty$. Therefore from (3.10) follows that there exists an integer $w_2 > w_1$ such that

$$m[T^{w_2}(A \cup T^{-w_1} A) \cap A] < \varepsilon_2.$$

We continue by induction. Having chosen the integers $w_0, w_1, \ldots, w_{k-1}$ we use (3.10) and choose an integer $w_k > w_{k-1}$ such that

$$m[T^{w_k} (A \cup T^{-w_1} A \cup \cdots \cup T^{-w_{k-1}} A) \cap A] < \varepsilon_k.$$

Let $A' = A \cap \left(\bigcup_{i=1}^{\infty} \bigcup_{j=0}^{i-1} T^{w_i - w_j} A \right)$ and $W = A \smallsetminus A'$.

Then $m(A') \leq \sum_{i=1}^{\infty} \varepsilon_i = \varepsilon$, and $m(W) \geq m(A) - \varepsilon > 0$.

We then have $W \subset A$, and for $i > j$ $T^{w_i-w_j} W \cap A \subset A'$. This says: for $i > j$ $T^{w_i-w_j} W \cap W = \emptyset$, or $T^{w_i} W \cap T^{w_j} W = \emptyset$. $\qquad \square$

Proof (Proposition 3.2.4).

Property (3.4) implies: $m(A) < \infty \implies \liminf_{n \to \infty} m(T^n A \cap A) = 0$. From Lemma 3.2.5 we conclude that T possesses a ww sequence.

Conversely, suppose C with $m(C) < \infty$, is a ww set with the sequence $\{w_i\}$, and let A and B be two sets of finite measure.

We first observe the following: for $p \in \mathbb{Z}$ and $n \in \mathbb{N}$

$$\left| \frac{1}{n} \sum_{i=0}^{n-1} m(T^{i+p} C \cap B) - \frac{1}{n} \sum_{i=0}^{n-1} m(T^i C \cap B) \right| = \frac{1}{n} \left| \sum_{i=p}^{p+n-1} m(T^i C \cap B) \right.$$

$$\left. - \sum_{i=0}^{n-1} m(T^i C \cap B) \right|$$

$$\leq \frac{2|p|}{n} m(B).$$

Then for any $r > 0$

$$\left| \frac{1}{n} \sum_{i=0}^{n-1} m\big(T^i [T^{w_1} C \cup \cdots \cup T^{w_r} C] \cap B\big) - r \frac{1}{n} \sum_{i=0}^{n-1} m(T^i C \cap B) \right|$$

$$= \left| \sum_{j=1}^{r} \frac{1}{n} \Big[\sum_{i=0}^{n-1} m(T^{i+w_j} C \cap B) - \sum_{i=0}^{n-1} m(T^i C \cap B) \Big] \right|$$

$$\leq \sum_{j=1}^{r} \frac{2|w_j|}{n} m(B) \longrightarrow 0 \quad \text{as } n \to 0.$$

Therefore

$$r \limsup_{n \to \infty} \frac{1}{n} \sum_{i=0}^{n-1} m(T^i C \cap B) = \limsup_{n \to \infty} \frac{1}{n} \sum_{i=0}^{n-1} m\big(T^i [T^{w_1} C \cup \cdots \cup T^{w_k} C] \cap B\big)$$

$$\leq m(B) < \infty.$$

Since $m(B) < \infty$ and $r > 0$ is arbitrary, the above says $\lim_{n \to \infty} \frac{1}{n} \sum_{i=0}^{n-1} m(T^i C \cap B) = 0$.

We note that for any integer $p \in \mathbb{Z}$ the set $T^p C$ is also a ww set for the same sequence $\{w_i\}$; therefore $\lim_{n \to \infty} \frac{1}{n} \sum_{i=0}^{n-1} m(T^i T^p C \cap B) = 0$.

Since $m(A) < \infty$, for any $\varepsilon > 0$ there exists an integer $k \in \mathbb{N}$ and a set D with $m(D) < \varepsilon$ such that $A = \left(\bigcup_{p=1}^{k} T^p C \cap A \right) \cup D$. Then

$$\lim_{n \to \infty} \frac{1}{n} \sum_{i=0}^{n-1} m(T^i A \cap B) \le \sum_{p=1}^{k} \left[\lim_{n \to \infty} \frac{1}{n} \sum_{i=0}^{n-1} m(T^i T^p C \cap B) \right]$$

$$+ \lim_{n \to \infty} \frac{1}{n} \sum_{i=0}^{n-1} m(T^i D \cap B)$$

$$\le \varepsilon.$$

This says $\displaystyle \lim_{n \to \infty} \frac{1}{n} \sum_{i=0}^{n-1} m(T^i A \cap B) = 0$.   □

3.3   Recurrent Sequences

Related to the *ww* sequences of a transformation there are other interesting sequences of integers that may (or may not) exist for some infinite ergodic transformations. We introduce the following:

Definition 3.3.1.   For a subset of integers $\mathbb{A} \subset \mathbb{Z}$ let us denote by $|\mathbb{A}|$ the number of elements in $\mathbb{A}$. Let T be an infinite ergodic transformation. An infinite set of integers $\{r_i\}$ is a *recurrent* sequence for the transformation T if $|\{r_i\} \cap \{w_i\}| < \infty$ for every *ww* sequence $\{w_i\}$ for T.

We show in Chap. 4, three important examples of infinite ergodic transformations. Among other things, the First and Second Basic Examples are infinite ergodic transformations that possess recurrent sequences, while the Third Basic Example is an ergodic transformation without recurrent sequences.

Analogous to the finite case, just as property (3.2) was strengthened to property (3.3), we strengthen property (3.4) for infinite ergodic transformations to:

$$m(A) < \infty, \ m(B) < \infty \implies \lim_{n \to \infty} m(T^n A \cap B) = 0. \qquad (3.11)$$

Finite ergodic transformations that satisfy property (3.3), namely the strongly mixing ones, have played a significant role in the classification of ergodic transformations in general. As a consequence some authors have been tempted to label infinite ergodic transformations that satisfy property (3.11) erroneously as "mixing" transformations in an infinite measure space. In what follows we show for infinite ergodic transformations both the existence and non-existence of recurrent sequences have interesting implications. In particular, in Theorems 3.3.11

and 3.3.12 we exhibit how "unmixing" an infinite ergodic transformation without recurrent sequences is.

The next proposition provides a description of recurrent sequences for a transformation without making reference to the ww sequences of the transformation.

Proposition 3.3.2. *Let T be an ergodic measure-preserving transformation defined on the infinite measure space $(X, \mathcal{B}, m)$. Then an infinite sequence of integers $\{r_i\}$ is a recurrent sequence for the transformation T if and only if there exists a set A of finite measure such that $\liminf\limits_{i\to\infty} m(T^{r_i} A \cap A) > 0$.*

Proof. If $\{r_i\}$ is not a recurrent sequence then it contains a ww subsequence $\{w_i\}$. Let C be a set of finite measure that is ww under the sequence $\{w_i\}$, and let A be any set of finite measure. The sequence $\{-w_i\}$ is also a ww sequence for the transformation T with the set C; therefore

$$m(A) \geq m\left(A \cap \bigcup_{i=0}^{\infty} T^{-w_i} C\right) = \sum_{i=0}^{\infty} m(A \cap T^{-w_i} C) = \sum_{i=0}^{\infty} m(T^{w_i} A \cap C).$$

This implies $\lim\limits_{i\to\infty} m(T^{w_i} A \cap C) = 0$.

We also have for any $j \in \mathbb{Z}$ the set $T^j C$ is a ww set with the same sequence $\{w_i\}$. This says $\lim\limits_{i\to\infty} m(T^{w_i} A \cap T^j C) = 0$.

Since $m(A) < \infty$, for any $\varepsilon > 0$ there exists an integer $N > 0$ and a set D with $m(D) < \varepsilon$ such that $A \subset \bigcup_{0 \leq j \leq N} (T^j C) \cup D$. Then

$$\lim_{i\to\infty} m\left(T^{w_i} A \cap A\right) = \lim_{i\to\infty} \left[m\left(T^{w_i} A \cap \bigcup_{j=0}^{N} T^j C\right) + m(T^{w_i} A \cap D) \right]$$

$$\leq \sum_{j=0}^{N} \lim_{i\to\infty} m\left(T^{w_i} A \cap T^j C\right) + \lim_{i\to\infty} m\left(T^{w_i} A \cap D\right) < \varepsilon,$$

and this implies $\lim\limits_{i\to\infty} m\left(T^{w_i} A \cap A\right) = 0$.

The converse implication follows from Lemma 3.2.5. $\square$

Remark 3.3.3. For an infinite ergodic transformation T defined on a σ-finite measure space $(X, \mathcal{B}, m)$, using Proposition 3.3.2, it is not difficult to show that

$$m(A) > 0 \quad \Longrightarrow \quad \limsup_{n\to\infty} m(T^n A \cap A) > 0 \tag{3.12}$$

is true for infinite ergodic transformations that possess recurrent sequences, while

$$m(A) < \infty \quad \Longrightarrow \quad \lim_{n\to\infty} m(T^n A \cap A) = 0 \tag{3.13}$$

is true for infinite ergodic transformations that do not possess recurrent sequences.

Corollary 3.3.4. *Let* T *be an ergodic measure-preserving transformation defined on the infinite measure space* $(X, \mathscr{B}, m)$. *Let* $\mathbb{U} = \{u_i\}$ *be an infinite subset of* $\mathbb{Z}$. *If* $\limsup_{i \to \infty} m(T^{u_i} A \cap A) > 0$ *for a set* A *with* $m(A) < \infty$ *then the sequence* $\mathbb{U}$ *contains a recurrent subsequence* $\mathbb{R} = \{r_i\}$.

Proof. Since $\limsup_{i \to \infty} m(T^{u_i} A \cap A) > 0$ for a set A of positive measure, the infinite sequence $\mathbb{U} = \{u_i\}$ contains a subsequence $\mathbb{R} = \{r_i\}$ with the property that $\liminf_{i \to \infty} m(T^{r_i} A \cap A) > 0$. Proposition 3.3.2 then completes the proof.   □

Corollary 3.3.5. *Let* T *be an ergodic measure-preserving transformation defined on the infinite measure space* $(X, \mathscr{B}, m)$. *Then property (3.11), namely*

$$m(A) < \infty, \quad m(B) < \infty \quad \Longrightarrow \quad \lim_{n \to \infty} m(T^n A \cap B) = 0,$$

is true for T *if and only if* T *does not possess recurrent sequences.*

Proof. Let A and B be two sets of finite measure. We note the inequality $m[T^n(A \cup B) \cap (A \cup B)] \geq m(T^n A \cap B)$ and apply Remark 3.3.3 to the set $A \cup B$.
   □

3.3.1   Transformations with Recurrent Sequences

The fact that recurrent sequences exist for some infinite ergodic transformation T is interesting in connection with the existence of a finite invariant measure as discussed in Chap. 1. In particular, it was shown by Y. Dowker [8] that a necessary and sufficient condition for the existence of a finite invariant measure $\mu \sim m$ was

$$\textbf{(D1)} \quad m(A) > 0 \quad \Longrightarrow \quad \liminf_{n \to \infty} m(T^n A) > 0.$$

This condition was weakened by A. Calderon [2] to

$$\textbf{(C)} \quad m(A) > 0 \quad \Longrightarrow \quad \liminf_{n \to \infty} \frac{1}{n} \sum_{i=1}^{n-1} m(T^i A) > 0$$

which in turn was weakened by Y. Dowker [9] to

$$\textbf{(D2)} \quad m(A) > 0 \quad \Longrightarrow \quad \limsup_{n \to \infty} \frac{1}{n} \sum_{i=1}^{n-1} m(T^i A) > 0.$$

We show that the existence of recurrent sequences for an infinite ergodic transformation precludes the further weakening of the above conditions for the existence of a finite invariant measure $\mu \sim m$. Namely, we show the following:

Proposition 3.3.6. *Let T be a measurable transformation defined on the finite measure space $(X, \mathscr{B}, m)$. Then the condition*

$$m(A) > 0 \implies \limsup_{n \to \infty} m(T^n A) > 0$$

is not a sufficient condition for the existence of a finite invariant measure $\mu \sim m$.

Proof. The proof follows from Lemmas 3.3.7–3.3.9.   □

Lemma 3.3.7. *Let T be a measurable transformation defined on the finite or σ-finite measure space $(X, \mathscr{B}, m_1)$, and let m_2 be a finite or σ-finite measure $m_2 \sim m_1$. Then*

$$\lim_{n \to \infty} m_1(T^n A \cap A) = 0 \quad \text{for some set } A \text{ with } m_1(A) > 0$$

if and only if

$$\lim_{n \to \infty} m_2(T^n B \cap B) = 0 \quad \text{for some set } B \text{ with } m_2(B) > 0.$$

Proof. For some set B of positive measure suppose $\lim_{n \to \infty} m_2(T^n B \cap B) = 0$. If $m_1(B) = \infty$ then B contains a subset A such that $0 < m_1(A) < \infty$. The set A still satisfies $\lim_{n \to \infty} m_2(T^n A \cap A) = 0$.

Let us consider the σ-field $\mathscr{A} = \{E : E \in \mathscr{B}, E \subset A\}$. If we restrict the measures m_1 and m_2 to the measurable space $(A, \mathscr{A})$ then Proposition 1.2.2 applies. Therefore given $\varepsilon > 0$ there exists a $\delta > 0$ such that $m_1(E) < \delta$ implies $m_2(E) < \varepsilon$ for $E \in \mathscr{A}$. From this follows $\lim_{n \to \infty} m_1(T^n A \cap A) = 0$.

Interchanging the measures m_1 and m_2 in the above completes the proof.   □

Lemma 3.3.8. *Let T be an infinite ergodic measure-preserving transformation defined on the σ-finite measure space $(X, \mathscr{B}, m)$. If $\lim_{n \to \infty} m(T^n B \cap B) = 0$ for some set B of positive measure then $\lim_{n \to \infty} m(T^n A \cap A) = 0$ for every set A of finite measure.*

Proof. Suppose $m(B) > 0$ and $\lim_{n \to \infty} m(T^n B \cap B) = 0$ for a set $B \in \mathscr{B}$. Then

$$\lim_{n \to \infty} m(T^{n+i} B \cap T^j B) = 0 \quad \text{for} \quad i, j \in \mathbb{Z}. \tag{3.14}$$

Since T is ergodic, for any set A of finite measure and for any $\varepsilon > 0$ there exist an integer $k > 0$ and a set C with $m(C) < \varepsilon$ such that $A \subset \bigcup_{i=0}^{k} T^i B \cup C$.

From the above and (3.14) follows $\limsup_{n \to \infty} m(T^n A \cap A) \leq 3\varepsilon$.   □

Lemma 3.3.9. *Let T be an ergodic transformation defined on the finite measure space $(X, \mathscr{B}, m)$. Then $\lim_{n \to \infty} m(T^n A \cap A) = 0$ if and only if $\lim_{n \to \infty} m(T^n A) = 0$.*

Proof. The if part is clear.

Conversely, suppose $\lim_{n\to\infty} m(T^n A \cap A) = 0$ for some set A with $m(A) > 0$. Since $m(X) < \infty$ and T is ergodic, we have for any $\varepsilon > 0$ an integer $k > 0$ and a set $C \in \mathscr{B}$ such that $X = \bigcup_{i=0}^{k} T^i A \cup C$ where $m(C) < \varepsilon$. Then

$$\limsup_{n\to\infty} m(T^n A) \leq \limsup_{n\to\infty} m\left[T^n A \cap \left(\bigcup_{i=0}^{k} T^i A \cup C \right) \right]$$

$$\leq \sum_{i=0}^{k} \limsup_{n\to\infty} m(T^n A \cap T^i A) + \limsup_{n\to\infty} m(T^n A \cap C)$$

$$\leq \varepsilon. \qquad\qquad\qquad \square$$

3.3.2 Transformations Without Recurrent Sequences

The fact that recurrent sequences do not exist for some infinite ergodic transformations is also interesting in connection with *ww* and *eww* sequences. In this case we describe sufficient conditions for an increasing sequence $\{n_i\}$ of integers to be a *ww* or *eww* sequence. We make the following definitions.

Definition 3.3.10. Let T be an infinite ergodic transformation defined on the σ-finite measure space $(X, \mathscr{B}, m)$.

- An increasing sequence $\{0 < N_1 < N_2 < \cdots\}$ of positive integers is a *ww growth* sequence for the transformation T if any sequence $\{0 < n_1 < n_2 < \cdots\}$ of positive integers that satisfies $n_i - n_{i-1} \geq N_i$ for $i \geq 1$ is a *ww* sequence for T.
- An increasing sequence $\{0 < N_1 < N_2 < \cdots\}$ of positive integers is an *eww growth* sequence for the transformation T if any sequence $\{0 < n_1 < n_2 < \cdots\}$ of positive integers that satisfies $n_i - 2n_{i-1} \geq N_i$ for $i \geq 1$ is an *eww* sequence for T.

Theorem 3.3.11. *Let T be an infinite ergodic transformation without recurrent sequences. Then there exists a ww growth sequence for the transformation T.*

Proof. Let $0 < m(A) < \infty$, and let

$$\lim_{n\to\infty} m(T^n A \cap A) = 0. \tag{3.15}$$

Let $0 < \varepsilon < m(A)$, and for $i = 1, 2, \ldots$ let $\varepsilon_i = \varepsilon/i2^i$. Using (3.15) we choose an increasing sequence of integers $0 < N_1 < N_2 < \cdots$ such that

$$\text{for each } i \geq 1 \quad m(T^n A \cap A) \leq \varepsilon_i \quad \text{for all } n \geq N_i. \tag{3.16}$$

We show that the sequence $\{N_i : i \geq 1\}$ is a *ww* growth sequence for T.

Let $0 < n_1 < n_2 < \cdots$ be an increasing sequence of positive integers satisfying $n_i - n_{i-1} \geq N_i$ for $i \geq 1$. We show $\{n_i\}$ is a ww sequence. Let $n_0 = 0$ and note that

$$\text{for } i > 0 \text{ and } 0 \leq j < i \text{ we have } n_i - n_j \geq n_i - n_{i-1} \geq N_i . \qquad (3.17)$$

Let $A' = \bigcup_{i=1}^{\infty} \bigcup_{j=0}^{i-1} T^{n_i - n_j} A \cap A$. In view of (3.16) and (3.17)

$$m(A') \leq \sum_{i=1}^{\infty} \sum_{j=0}^{i-1} m(T^{n_i - n_j} A \cap A) \leq \sum_{i=1}^{\infty} i \varepsilon_i = \sum_{i=1}^{\infty} \frac{\varepsilon}{2^i} = \varepsilon .$$

Let $A_0 = A \smallsetminus A'$. Then $m(A_0) \geq m(A) - \varepsilon > 0$. We show

$$T^{n_i} A_0 \cap T^{n_j} A_0 = \emptyset \quad \text{for} \quad i, j \geq 0 \quad \text{and} \quad i \neq j. \qquad (3.18)$$

We assume $i > j$. Then, since $A_0 = A \setminus A'$ and $T^{n_i - n_j} A_0 \subset A'$, we have

$$T^{n_i - n_j} A_0 \cap A_0 = \emptyset \quad \text{for} \quad i, j \geq 0.$$

This implies (3.18) and proves $\{n_i\}$ is a ww sequence for T.   □

Theorem 3.3.12. *Let T be an infinite ergodic transformation without recurrent sequences. Then there exists an eww growth sequence for the transformation T.*

Proof. Let $0 < m(A) < \infty$, and let

$$\lim_{n \to \infty} m(T^n A \cap A) = 0. \qquad (3.19)$$

Let $0 < \varepsilon < m(A)$, and for $i \geq 1$ let $\varepsilon_i = \dfrac{\varepsilon}{i^3 (2i + 1) 3^3 2^{i+1}}$.
Using (3.19) we choose an increasing sequence of integers $0 < N_1' < N_2' < \cdots$ such that

$$\text{for each } i \geq 1 \quad m(T^n A \cap A) \leq \varepsilon_i \quad \text{for all } n \geq N_i'. \qquad (3.20)$$

We let $N_i = N_i' + i$, and let $0 = n_0 < n_1 < \cdots$ be an increasing sequence of integers satisfying $n_i - 2n_{i-1} \geq N_i$ for $i \geq 1$. We show $\{n_i : i > 0\}$ is an sww sequence for T, and this proves $\{N_i\}$ is an eww growth sequence for it.

For each $i = 1, 2, \ldots$ we consider the set of integers
$S_i = \{s : s = an_i + bn_j + cn_k + dn_l + e\}$;   where
$a \in \{1, 2\}$ (and note there are at most 2 choices for a),
$b, c, d \in \{0, \pm 1\}$ (and note there are at most 3^3 choices for $\{b, c, d\}$),

$e \in \{0, \pm 1, \ldots \pm i\}$ (and note there are at most $2i + 1$ choices for e),
$0 \le j, k, l < i$ (and note there are at most i^3 choices for $\{j, k, l\}$).
From this follows that $|S_i| \le 2i^3 3^3 (2i + 1)$.

In the choice of the sets S_i we also require that at most two of the numbers b, c, d be negative.

Since $\{n_i : i > 0\}$ is an increasing sequence of positive integers it follows that for $i = 1, 2, \ldots$ and $s \in S_i$

$$s = an_i + bn_j + cn_k + dn_l + e \ge n_i - 2n_{i-1} - i \ge N_i - i = N_i'. \quad (3.21)$$

Let $A' = \bigcup_{i=1}^{\infty} \bigcup_{s \in S_i} T^s A \cap A$. Then in view of (3.20) and (3.21)

$$m(A') \le \sum_{i=1}^{\infty} \sum_{s \in S_i} m(T^s A \cap A) \le \sum_{i=1}^{\infty} |S_i| \varepsilon_i$$

$$= \sum_{i=1}^{\infty} 2i^3 3^3 (2i + 1) \varepsilon_i = \sum_{i=1}^{\infty} \frac{\varepsilon}{2^i} = \varepsilon.$$

Let $A_0 = A \smallsetminus A'$. Then $m(A_0) \ge m(A) - \varepsilon > 0$.

We show for $i, j, k, l \ge 0$ and $i > j$

$$T^{n_i - n_k + k} A_0 \cap T^{n_j - n_l + l} A_0 = \emptyset \quad (3.22)$$

whenever one of the indices $\{i, j, k, l\}$ is larger than all the rest or $i = l > \max\{j, k\}$.

In the case where one of the indices $\{i, j, k, l\}$, let us say k, is larger than all the rest then we have:

$$s = n_k - n_i + n_j - n_l - k + l \in S_k, \quad A_0 = A \setminus A', \quad T^s A_0 \subset A', \quad T^s A_0 \cap A_0 = \emptyset.$$

This implies (3.22). A similar proof applies for the other indices in this case.

In the case where $i = l > \max\{j, k\}$:

$$s = 2n_i - n_k - k - n_j + i \in S_i, \quad A_0 \subset A, \quad T^s A_0 \subset A', \quad T^s A_0 \cap A_0 = \emptyset.$$

Again this implies (3.22), which shows $\{n_i : i \ge 0\}$ is an *sww* sequence for T. Proposition 2.2.2 then implies $\{n_i : i \ge 0\}$ is an *eww* sequence for T. $\square$

We summarize the above discussion in the following theorem.

Theorem 3.3.13 (Infinite Ergodic Transformations). *An infinite ergodic transformation T defined on the measure space $(X, \mathcal{B}, m)$ has the following properties:*

(T1) *T does not preserve a finite measure $\mu \sim m$.*

(T2) $m(A) < \infty, \ m(B) < \infty \implies \lim_{n \to \infty} \frac{1}{n} \sum_{i=0}^{n-1} m(T^i A \cap B) = 0$.

(T3) T *is a transformation without recurrent sequences if and only if*
$m(A) < \infty, \ m(B) < \infty \implies \lim_{n \to \infty} m(T^n A \cap B) = 0$.

(T4) T *has no wandering sets.*

(T5) T *has weakly wandering sets of infinite measure.*

(T6) *Every sww sequence for* T *is an eww sequence for* T.

(T7) T *has hereditary eww sequences.*

(T8) *When* T *has no recurrent sequences*
it possesses both ww and eww growth sequences.

Chapter 4
Three Basic Examples

In this chapter we present and discuss three basic examples of infinite ergodic transformations. We show some special and unique properties these transformations possess. These properties involve characteristics of infinite ergodic transformations that distinguish the transformations from finite ergodic transformations.

In the First Basic Example we construct an infinite ergodic transformation that possesses recurrent sequences. Moreover, we show that it possesses an *eww* set of finite measure. The Second Basic Example is an infinite ergodic transformation that also has recurrent sequences; however, it does not have any *eww* sets of finite measure. Finally, the Third Basic Example is an infinite ergodic transformation without any recurrent sequences.

4.1 First Basic Example

There are many examples of infinite ergodic transformations in the literature. However, the existence of *ww* sets was not suspected for any of them until it was shown in [32] that every infinite ergodic transformation possesses *ww* sets. Nevertheless, when attempts were made to exhibit an explicit example of a *ww* set for any of the known infinite ergodic transformations we could not find a concrete example of a *ww* set for any of them. The transformation T of the First Basic Example that follows was then constructed with the specific purpose of exhibiting a *ww* set for it. As we shall see, the transformation T turned out to possess yet a few more unexpected and interesting properties.

© Springer Japan 2014
S. Eigen et al., *Weakly Wandering Sequences in Ergodic Theory*,
Springer Monographs in Mathematics, DOI 10.1007/978-4-431-55108-9_4

4.1.1   Induced Transformations

For the construction of the transformation of the First Basic Example we discuss a
general method of building infinite ergodic transformations using induced transformations. This method was introduced and discussed in [43].

Let T be an ergodic measure-preserving transformation defined on the σ-finite
measure space $(X, \mathcal{B}, m)$, and let $X_0 \subset X$ be a subset of positive measure with
$m(X_0) < \infty$. After possibly removing from X_0 a subset of measure zero, every
point $x \in X_0$ returns to X_0 infinitely often under positive and negative iterates
of T. Let us denote by $\mathcal{B}_0$ the collection $\{B \in \mathcal{B} : B \subset X_0\}$ and restrict the
measure m on the measurable space $(X_0, \mathcal{B}_0)$ and denote it by m_0. Then the *induced
transformation* S on the measure space $(X_0, \mathcal{B}_0, m_0)$ is defined by $Sx = T^n x$,
where $n > 0$ is the smallest positive integer such that $T^n x \in X_0$. It is also useful to
describe the transformation S on $(X_0, \mathcal{B}_0, m_0)$ as follows:
For $n \geq 1$ let $A_n = \{x \in X_0 : T^n x \in X_0,$ and $T^k x \notin X_0$ for $0 < k < n\}$. Then for
$x \in A_n$ and $n \geq 0$ $Sx = T^n x$. It follows that S is an ergodic measure-preserving
transformation defined on the finite measure space $(X_0, \mathcal{B}_0, m_0)$.

It is possible to go in the other direction as well. Let us start with an
ergodic measure-preserving transformation S defined on a finite measure space
$(X_0, \mathcal{B}_0, m_0)$. Let $\{A_n : n \geq 1\}$ be a sequence of mutually disjoint subsets of X_0
such that $X_0 = \bigcup_{n=1}^{\infty} A_n$ *(disj)*. Since we are assuming that $m_0(X_0) < \infty$ we have
$m(A_n) \to 0$ as $n \to \infty$.

Next we choose a sequence of nonnegative integers

$$\{f(n) : 0 \leq f(0) \leq f(1) \leq f(2) \leq \cdots\}.$$

For each $n = 0, 1, 2, \ldots$ if $f(n) > 0$ then consider $f(n) + 1$ mutually disjoint and
isomorphic copies of A_n; namely $A_n^0 = A_n, A_n^1, \ldots, A_n^{f(n)}$.
Let us denote by the same letter R all the isomorphisms

$$RA_n^i = A_n^{i+1} \quad \text{for } i = 0, 1, \ldots, f(n) - 1 \text{ and } n \geq 1.$$

Next we consider the measure space $(X, \mathcal{B}, m)$ where

$$X = \bigcup_{n=0}^{\infty} \bigcup_{i=0}^{f(n)} A_n^i \ (disj),$$

$\mathcal{B}$ is the σ-field of all measurable subsets of X, and m the corresponding measure
on $(X, \mathcal{B})$. The transformation $T : X \to X$ is then defined by

$$Tx = \begin{cases} Rx & \text{if } x \in A_n^i, \ i = 0, 1, \ldots, f(n) - 1; \ n \geq 0, \\ S\left(R^{-f(n)}x\right) & \text{if } x \in A_n^{f(n)}, \ n \geq 0. \end{cases}$$

In the above construction it follows that T is an ergodic measure-preserving transformation on $(X, \mathscr{B}, m)$ where

$$m(X) = \sum_{n=0}^{\infty} (f(n) + 1) \, m_0(A_n) \,.$$

In particular, we may start with any ergodic measure-preserving transformation S on a finite measure space $(X_0, \mathscr{B}_0, m_0)$, and choose a sequence of subsets $\{A_n : n \geq 1\}$ such that $\sum_{n=0}^{\infty} (f(n) + 1) \, m(A_n) = \infty$. Then we obtain an ergodic measure-preserving transformation T on $(X, \mathscr{B}, m)$ with $m(X) = \infty$.

4.1.2 Construction of the First Basic Example

Using the above we construct our First Basic Example of an infinite ergodic transformation T; see [33].

Let $(X_0, \mathscr{B}_0, m_0)$ be a finite measure space where

$$X_0 = \{x \in \mathbb{R}, \ 0 < x < 1, \ x \neq \text{dyadic rational}\} \,,$$

$\mathscr{B}_0$ is the σ-field of all Lebesgue measurable subsets of X_0 and m_0 is the ordinary Lebesgue measure on $\mathscr{B}_0$ with $m_0(X_0) = 1$.

For $n \geq 0$ we let

$$A_n = \{x : x \in X_0, \ 1 - 1/2^n < x < 1 - 1/2^{n+1}\}$$

be the dyadic interval of length $1/2^{n+1}$. We define the transformation S on X_0 onto itself as follows:

$$Sx = x - (1 - 3/2^{n+1}) \quad if \ \ x \in A_n, \ \ n \geq 0 \,.$$

Then S is an ergodic measure-preserving transformation defined on the measure space $(X_0, \mathscr{B}_0, m_0)$ with $m_0(X_0) = 1$. This transformation is usually referred to as the von Neumann adding machine transformation (or the dyadic odometer) (Fig. 4.1).

Let us put $f(0) = 0$ and

$$f(n) = 2 + 2^3 + \cdots + 2^{2n-1} \quad \text{for } n \geq 1 \,.$$

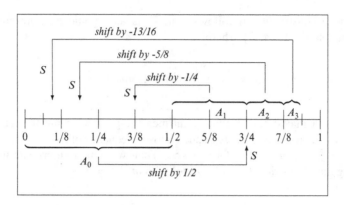

Fig. 4.1 The transformation S on the space X_0

For $n \geq 1$ we consider $f(n) + 1$ mutually disjoint copies of A_n; namely $A_n^0 = A_n$, A_n^1, ..., $A_n^{f(n)}$. We denote by the same letter R, all the following isomorphisms:

$$RA_n^i = A_n^{i+1}, \quad i = 0, 1, \ldots, f(n) - 1; \; n \geq 1 \,.$$

Next we consider the measure space $(X, \mathscr{B}, m)$ where

$$X = \bigcup_{n=0}^{\infty} \bigcup_{i=0}^{f(n)} A_n^i (disj) \,,$$

$\mathscr{B}$ is the σ-field of all "Lebesgue measurable" subsets of X, and m is the corresponding measure on $(X, \mathscr{B})$.

We now define the transformation T from X onto itself as follows:

$$Tx = \begin{cases} Rx & \text{if } x \in A_n^i \; for \; i = 0, 1, \ldots, f(n) - 1 \; and \; n \geq 0, \\ SR^{-f(n)}x & \text{if } x \in A_n^{f(n)} \; for \; n \geq 0 \,. \end{cases} \quad (4.1)$$

It follows that T is an ergodic measure-preserving transformation defined on the σ-finite measure space $(X, \mathscr{B}, m)$ with $m(X) = \infty$. This T is the First Basic Example and a diagram of its action on X is given in Fig. 4.2.

We introduce the following subsets of X: $\quad B_0 = X_0$,

$$B_j = \bigcup_{n=j}^{\infty} \bigcup_{i=f(j-1)+1}^{f(j)} A_n^i (disj) \quad \text{for } j \geq 1 \quad \text{and}$$

$$C_k = \bigcup_{j=0}^{k} B_j (disj) \quad \text{for } k \geq 0 \,.$$

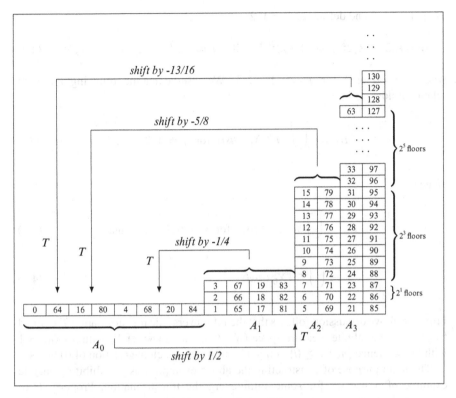

Fig. 4.2 The First Basic Example T acting on the space X

We observe that the sets B_j and C_k have the following properties:

$$m(B_j) = [f(j) - f(j-1)] \sum_{n=j}^{\infty} 1/2^{n+1} = 2^{j-1} \quad \text{for } j \geq 1,$$

$$m(C_k) = 1 + \sum_{j=1}^{k} m(B_j) = 2^k \quad \text{for } k \geq 0,$$

$$X = \bigcup_{j=0}^{\infty} B_j \ (disj), \quad \text{and}$$

$$C_0 \subset C_1 \subset C_2 \subset \cdots \subset C_k \subset \cdots \ \nearrow \ X. \tag{4.2}$$

Also from the above definition (4.1) of T we get

$$T^n(C_k) \subset C_{k+l} \quad \text{for } n = 0, 1, \ldots, f(k+l) - f(k), \ k \geq 0, \ l \geq 0. \tag{4.3}$$

We put $n_0 = 0$ and define for $i = 1, 2, \ldots$

$$n_i = \varepsilon_0 2^1 + \varepsilon_1 2^3 + \cdots + \varepsilon_k 2^{2k+1} \quad \text{if} \quad i = \varepsilon_0 2^0 + \varepsilon_1 2^1 + \cdots + \varepsilon_k 2^k, \quad (4.4)$$

where $\varepsilon_j = 0$ or 1 for $j = 0, 1, \ldots, k$. We observe that the following important relation holds:

$$B_j = \bigcup_{i=2^{j-1}}^{2^j - 1} T^{n_i} X_0 \ (disj) \ \text{for} \ j = 1, 2, \ldots .$$

Hence

$$C_k = \bigcup_{i=0}^{2^k - 1} T^{n_i} X_0 \ (disj) \ \text{for} \ k = 0, 1, \ldots , \ \text{and} \qquad (4.5)$$

$$X = \bigcup_{i=0}^{\infty} T^{n_i} X_0 \ (disj) . \qquad (4.6)$$

From the above discussion follows that the infinite ergodic transformation T defined by (4.1) on the infinite measure space $(X, \mathcal{B}, m)$ has a *ww* set X_0 with $m(X_0) = 1$ with the sequence $\{n_i : i \geq 0\}$ defined by (4.4) for which the relation (4.6) holds.

The main purpose of constructing the above example was to exhibit a concrete example of a *ww* set for some infinite ergodic transformation. Property (4.6) accomplished that; however, Property (4.6) exhibited yet another surprising and interesting property of the transformation T. It showed that the sequence $\{n_i : i \geq 0\}$ described above is in fact an *eww* sequence for the transformation T with the set X_0 with the additional property that $m(X_0) < \infty$. While studying various necessary and sufficient conditions for a finite invariant measure, it was unexpected and surprising to realize that infinite ergodic transformations always possessed *ww* sequences. The fact that an infinite ergodic transformation may possess *eww* sequences was equally unexpected and surprising. It was in this First Basic Example that *eww* sequences first appeared, and the fact that there existed an *eww* sequence with an *eww* set of finite measure was a bigger surprise. This property of an infinite ergodic transformation possessing an *eww* sequence with an *eww* set of finite measure proved to have some important and significant consequences as well. We discuss this in more detail in Chap. 5.

It is clear by now that an infinite ergodic transformation possesses *ww* sequences. It is a difficult task, however, to determine all the *ww* sequences for any infinite ergodic transformation. The case for recurrent sequences is somewhat different. Recurrent sequences for an infinite ergodic transformation were defined and discussed in Chap. 3. As we shall see soon, the infinite ergodic transformation T of the Second Basic Example that we present next, possesses recurrent sequences, while the Third Basic Example that follows, is an infinite ergodic transformation without recurrent sequences. The infinite ergodic transformation of the First Basic Example also possesses recurrent sequences. In particular, it is not difficult to see that the sequence {all finite sums of even powers of 2} is a recurrent sequence for it. For the transformation T of the First Basic Example we can do more. In the next theorem we give a complete description of all recurrent sequences for this transformation.

Let us consider the following sequence of integers: let $N_0 = \{0\}$, and

$$N_k = \{n : n = \pm 2^{2p_1} \pm 2^{2p_2} \pm \cdots \pm 2^{2p_k}\} \text{ for } k \geq 1 \qquad (4.7)$$

where $p_1, p_2, \ldots, p_k$ are integers satisfying $0 \leq p_1 < p_2 < \cdots < p_k$. It is not difficult to show that the set X_0 defined above satisfies:

$$m(T^n X_0 \cap X_0) = \begin{cases} 1/2^k & \text{if } n \in N_k, \ k = 0, 1, 2, \ldots, \\ 0 & \text{if } n \in \mathbb{Z} \smallsetminus \bigcup_{k=0}^{\infty} N_k. \end{cases} \qquad (4.8)$$

Theorem 4.1.1. *The infinite set of integers $\{r_i : i = 0, 1, 2, \ldots\}$ is a recurrent sequence for the infinite ergodic transformation T of the First Basic Example if and only if there exist two positive integers l and s such that*

$$\{r_i : i = 0, 1, 2, \ldots\} \subset \bigcup_{n=-s}^{s} \bigcup_{k=0}^{l} (N_k + n) \qquad (4.9)$$

where $\{N_k : k = 0, 1, 2, \ldots\}$ is the sequence described in (4.7).

Proof. Let $\{r_i : i = 0, 1, 2, \ldots\}$ be a recurrent sequence for T, and let A be a set of finite measure such that

$$\liminf_{i \to \infty} m(T^{r_i} A \cap A) > 3\delta > 0.$$

From (4.2) follows that there exists a positive integer k such that $m(A \smallsetminus C_k) < \delta$. Then we have

$$\liminf_{i \to \infty} m(T^{r_i} C_k \cap C_k) > \delta > 0.$$

From (4.5) follows

$$\lim_{i \to \infty} \sum_{p=0}^{2^k-1} \sum_{q=0}^{2^k-1} m(T^{r_i + n_p - n_q} X_0 \cap X_0) > \delta > 0.$$

Let l be a positive integer such that $1/2^l < \delta/2^{2k}$. Then there exists a positive integer i_0 such that for any $i > i_0$, there exist two integers p and q ($0 \leq p, q \leq 2^k - 1$) such that

$$m(T^{r_i + n_p - n_q} X_0 \cap X_0) \geq 1/2^l. \tag{4.10}$$

We note that (4.10) is equivalent to saying that

$$r_i + n_p - n_q \in \bigcup_{k=0}^{l} N_k, \quad \text{or} \quad r_i \in \bigcup_{k=0}^{l} (N_k - n_p + n_q).$$

This shows the necessity of (4.9). The sufficiency is obvious.   □

This example has a property sharper than just possessing recurrent sequences. It is possible to show that it also possesses the following "regular" behavior. Namely, $m(A) > 0 \implies \limsup_{n \to \infty} m(T^n A \cap A) = \frac{1}{2} m(A)$.

4.2 Second Basic Example

As we saw above, the infinite ergodic transformation T of the First Basic Example had the following interesting property:

T has an *eww* set of finite measure.

It is not difficult to see that the above property in turn implies that the centralizer of the transformation T consists of only measure-preserving transformations (see Corollary 5.1.4). In other words, if $TQ = QT$ for some measurable transformation Q, then Q preserves the same invariant measure m for T. Initially, it was not known whether all infinite ergodic transformations possessed the above property. The infinite ergodic transformation T that we construct in the Second Basic Example does not have that property. In other words, the centralizer of the Second Basic Example contains a transformation Q which does not preserve the T-invariant measure m. From this it follows that every *eww* set for the Second Basic Example T must be of infinite measure.

4.2.1 Non-measure-Preserving Commutators

Before constructing the Second Basic Example, let us consider an infinite ergodic transformation T defined on the σ-finite measure space $(X, \mathscr{B}, m)$ that commutes with a transformation Q which is not measure-preserving. Using the pair of transformations T and Q we show how to construct in a systematic way an ergodic transformation R that does not preserve any σ-finite measure $\mu \sim m$; in other words, a transformation R which is an ergodic transformation of type III.

First we prove the following lemma and make a few observations.

Lemma 4.2.1. *If a measurable and nonsingular transformation Q has the property that for a constant $\alpha \neq 1$, $m(QA) = \alpha m(A)$ for all $A \in \mathcal{B}$, then there exists a wandering set W for Q satisfying $X = \bigcup_{n \in \mathbb{Z}} Q^n W$ (disj). In other words, Q is a dissipative transformation.*

Proof. We assume $0 < \alpha < 1$; otherwise we replace Q by Q^{-1}. For any set $B \in \mathcal{B}$ with $0 < m(B) < \infty$ and $n \in \mathbb{N}$ we let

$$B_n = \{x \in B : Q^n x \in B \text{ and } Q^i x \notin B \text{ for } 1 \leq i < n\}.$$

It follows that

$$B \supset \bigcup_{n \in \mathbb{N}} B_n (disj) \quad \text{and} \quad B \supset \bigcup_{n \in \mathbb{N}} Q^n B_n (disj). \tag{4.11}$$

We let

$$W_1 = B \smallsetminus \bigcup_{n \in \mathbb{N}} Q^n B_n = \{x \in B : Q^{-i} x \notin B \text{ for all } i \in \mathbb{N}\}. \tag{4.12}$$

Then from (4.11) and the fact that $\alpha < 1$ we have

$$m(W_1) = m(B) - \sum_{n=1}^{\infty} m(Q^n B_n) = m(B) - \sum_{n=1}^{\infty} \alpha^n m(B_n) > 0.$$

We conclude from (4.12) that

$$Q^i W_1 \cap Q^j W_1 = \emptyset \quad \text{for} \quad i, j \in \mathbb{Z} \quad \text{and} \quad i \neq j.$$

Next we let

$$W_1^* = \bigcup_{n \in \mathbb{Z}} Q^n W_1 (disj).$$

In the case where $m(X \smallsetminus W_1^*) > 0$ we repeat the above argument and obtain a set $W_2 \subset X \smallsetminus W_1^*$ that satisfies:

$$m(W_2) > 0 \quad \text{and} \quad W_2^* = \bigcup_{n \in \mathbb{Z}} Q^n W_2 (disj).$$

It follows that the set $W_1 \cup W_2$ is again a wandering set for Q.

We continue this way, by transfinite induction if necessary, and finally obtain a countable number of sets $W_1, W_2, \ldots$, satisfying:

$$X = \bigcup_{n \in \mathbb{Z}} Q^n W \ (disj) \quad \text{where} \quad W = \bigcup_{n \in \mathbb{N}} W_n.$$

$\square$

Suppose that T is an infinite ergodic transformation and Q is a dissipative transformation with the property that $TQ = QT$. It follows that the transformations T and Q also satisfy:

For all $x \in X$ the transformation T maps $Orb_Q(x)$ onto $Orb_Q(Tx)$, (4.13)

where $Orb_Q(x) = \{Q^n x : n \in \mathbb{Z}\}$ is the Q-orbit of x.

For the purpose of Theorem 4.2.2 that follows, let us consider an infinite ergodic transformation T and a dissipative transformation Q defined on the σ-finite measure space $(X, \mathcal{B}, m)$ that satisfy (4.13). Then $X = \bigcup_{n \in \mathbb{Z}} Q^n Y$ where Y is a wandering set for Q. We note that for all $x \in X$ the point $x' = Orb_Q(x) \cap Y$ is uniquely defined. Next we consider the measure space $(Y, \mathcal{B}_Y, m)$ where $\mathcal{B}_Y = \{B \subset Y : B \in \mathcal{B}\}$ and m is the restriction on $(Y, \mathcal{B}_Y)$ of the measure m defined on $(X, \mathcal{B})$. We define the transformation R on Y as follows:

$$Ry = (Ty)' = Orb_Q(Ty) \cap Y \quad \text{for} \quad y \in Y. \qquad (4.14)$$

Theorem 4.2.2. *Let T be a measure-preserving ergodic transformation and Q a dissipative transformation defined on the σ-finite measure space $(X, \mathcal{B}, m)$ satisfying: T maps $Orb_Q(x)$ onto $Orb_Q(Tx)$ for all $x \in X$. Then the transformation R as defined in (4.14) is an ergodic transformation that does not preserve any σ-finite measure $\mu \sim m$ if and only if Q is not measure-preserving.*

Proof. We show that R is an ergodic transformation, and in the case where Q does not preserve the measure m then there does not exist a measure $\mu \sim m$ invariant for R; the rest is clear.

Let $B \in \mathcal{B}_Y$ with $RB = B$, and let $B^* = \bigcup_{n \in \mathbb{Z}} Q^n B$. From the definition of R and the fact that T maps $Orb_Q(x)$ onto $Orb_Q(Tx)$ follows: $TB^* = B^*$. Then T ergodic implies that either $m(B^*) = 0$ or $m(X \smallsetminus B^*) = 0$, which implies either $m(B) = 0$ or $m(Y \smallsetminus B) = 0$. This proves that R is ergodic.

Next we assume that m is not an invariant measure for Q and assume there exists a measure $\mu \sim m$ which is invariant for R. We extend the measure μ to the measurable space $(X, \mathcal{B})$, denote it by the same symbol μ, and define it as follows: For any set $B \in \mathcal{B}$ and $B \subset Q^n Y$ for some $n \in \mathbb{Z}$ we let $\mu(B) = \mu(Q^{-n} B)$. The nonsingularity of Q implies $\mu \sim m$ on the measurable space $(X, \mathcal{B})$, and from the definition of μ on $(X, \mathcal{B})$ it follows that $\mu(QB) = \mu(B)$ for all $B \in \mathcal{B}$.

We show that μ is an invariant measure for T. We let $B \in \mathscr{B}$, and for $i, j, k \in \mathbb{Z}$ we let

$$B_{i,j,k} = B \cap Q^i Y \cap T^{-1} Q^j Y \cap Q^i T^{-1} Q^k Y. \tag{4.15}$$

Clearly (4.15) implies

$$B_{i,j,k} \subset Q^i Y, \tag{4.16}$$

$$T B_{i,j,k} \subset Q^j Y, \tag{4.17}$$

and

$$T Q^{-i} B_{i,j,k} \subset Q^k Y. \tag{4.18}$$

Since for all $x \in X$ the transformation T maps $Orb_Q(x)$ onto $Orb_Q(Tx)$, and for any $z \in Orb_Q(x)$ there exists a unique integer p such that $Q^p z \in Y$, from (4.17) and (4.18) follows

$$Q^{-j} T B_{i,j,k} = Q^{-k} T Q^{-i} B_{i,j,k} \subset Y. \tag{4.19}$$

From (4.17), (4.19), the definition of R and the fact that μ is an invariant measure for both R and Q we get

$$\mu(T B_{i,j,k}) = \mu(Q^{-j} T B_{i,j,k}) = \mu(Q^{-k} T Q^{-i} B_{i,j,k}) \tag{4.20}$$

$$= \mu(R Q^{-i} B_{i,j,k}) = \mu(Q^{-i} B_{i,j,k}) = \mu(B_{i,j,k}).$$

We also note that the set $B \in \mathscr{B}$ can be expressed as

$$B = \bigcup_{i,j,k \in \mathbb{Z}} B_{i,j,k} \; (disj), \tag{4.21}$$

From (4.20) and (4.21) we conclude that μ defined on $(X, \mathscr{B})$ is an invariant measure for T. Proposition 2.2.1 **(5E)** then implies that there exists a constant $\alpha > 0$ such that

$$\mu(B) = \alpha m(B) \quad \text{for all} \quad B \in \mathscr{B}. \tag{4.22}$$

The measure m is not an invariant measure for Q. Therefore

$$m(QB) \neq m(B) \quad \text{for some} \quad B \in \mathscr{B}. \tag{4.23}$$

Combining (4.22), (4.23) and using the fact that μ is an invariant measure for Q we get

$$\mu(B) = \mu(QB) = \alpha m(QB) \neq \alpha m(B) = \mu(B) \quad \text{for some} \quad B \in \mathscr{B}.$$

This is a contradiction.   □

4.2.2   A General Class of Transformations

Let us consider the one-sided infinite direct product measure space

$$(Y, \mathscr{B}_Y, m) = \prod_{i=1}^{\infty} (Y_i, \mathscr{B}_i, m_i),$$

where $Y_i = \{0, 1, \ldots, k\}$, $\mathscr{B}_i =$ all subsets of Y_i, and $m_i(0) = p_0$, $m_i(1) = p_1, \ldots, m_i(k) = p_k$ with $p_j > 0$ for j satisfying $0 \leq j \leq k$ and $\sum_{j=0}^{k} p_j = 1$.

Since $p_j > 0$ for each j ($0 \leq j \leq k$) it is clear that the set N consisting of all those $y = (\varepsilon_1, \varepsilon_2, \ldots) \in Y$ for which there are only finitely many i's satisfying $\varepsilon_i = j$ for some j ($0 \leq j \leq k$) satisfies $m(N) = 0$. We remove this set from Y in the sequel. We then define on $(Y, \mathscr{B}_Y, m)$ the transformation R, at times called the *odometer* or *adding machine* transformation, as follows:

For each $n \in \mathbb{N}$ and j satisfying $0 \leq j \leq k - 1$ we consider the cylinder set

$$Y[n, j] = \{y = (\varepsilon_1, \varepsilon_2, \ldots) \in Y : \varepsilon_1 = k, \ldots, \varepsilon_{n-1} = k, \varepsilon_n = j\}.$$

Then for $n \in \mathbb{N}$, j with $0 \leq j \leq k - 1$ and $y = (k, \ldots, k, j, \varepsilon_{n+1}, \varepsilon_{n+2}, \ldots)$ we define

$$Ry = (0, \ldots, 0, j + 1, \varepsilon_{n+1}, \varepsilon_{n+2}, \ldots).$$

It is clear that $Y = \bigcup_{n \in \mathbb{N}} \bigcup_{0 \leq j \leq k-1} Y[n, j]$, and $R : Y \longrightarrow Y$ is a measurable and nonsingular transformation defined on the measure space $(Y, \mathscr{B}_Y, m)$. Furthermore, the transformation R preserves the measure m if and only if $p_0 = p_1 = \cdots = p_k = \frac{1}{k+1}$. In particular, if $k = 1$ and $p_0 = p_1 = \frac{1}{2}$ then the corresponding odometer transformation preserves the measure m, and it is not difficult to verify that the mapping $\Theta : Y \longrightarrow [0, 1]$ defined by

$$y = (\varepsilon_1, \varepsilon_2, \ldots) \longmapsto \Theta(y) = \sum_{i=1}^{\infty} \frac{\varepsilon_i}{2^i}$$

gives a measure-theoretic isomorphism between this odometer transformation and the von Neumann transformation discussed in the First Basic Example above.

We show next that every transformation R defined as above is an ergodic transformation, and R does not preserve any measure equivalent to m except for the case when $p_0 = p_1 = \cdots = p_k = \frac{1}{k+1}$.

For each $n \in \mathbb{N}$ let $\mathscr{G}_n$ be the group of permutations of the elements of the set $\{1, 2, \ldots, n\}$. Each element π of the group $\mathscr{G}_n$ defines a measurable transformation Ψ_π on the product measure space $(Y, \mathscr{B}_Y, m)$ given by

$$y = (\varepsilon_1, \ldots, \varepsilon_n, \varepsilon_{n+1}, \ldots) \longmapsto \Psi_\pi(y) = (\varepsilon_{\pi(1)}, \ldots, \varepsilon_{\pi(n)}, \varepsilon_{n+1}, \ldots).$$

It is easy to see that for each $y \in Y$ and $q \in \mathbb{Z}$, $\Psi_\pi^q(y)$ belongs to the set $Orb_R(y)$, where R is the odometer transformation on the space $(Y, \mathscr{B}_Y, m)$ introduced above. If we define $\mathscr{G} = \bigcup_{n \in \mathbb{N}} \mathscr{G}_n$ then $\mathscr{G}$ can be regarded in a natural way as a group of nonsingular measurable transformations on $(Y, \mathscr{B}_Y, m)$. Then the observation above tells us that it is a subgroup of the full group $[R]$ where

$$[R] = \{U : Y \longrightarrow Y \text{ and } Orb_U(y) \subset Orb_R(y) \text{ for all } y \in Y\}.$$

There is also another transformation, let us call it S, which belongs to the full group $[R]$ and plays a significant role in our arguments in the sequel. In order to define S we need more terminology.

For each j in the set $\{0, 1, \ldots, k\}$ denote by χ_j the indicator function of the set $\{j\}$; namely

$$\chi_j(i) = \begin{cases} 1 & \text{if } j = i, \\ 0 & \text{if } j \neq i. \end{cases}$$

For each $q \in \mathbb{N}$ and a point $y = (\varepsilon_1, \varepsilon_2, \ldots)$ the transformation R^q changes only a finite number of coordinates ε_i's. Let us write $R^q(y) = (\varepsilon_1', \varepsilon_2', \ldots)$. Then the quantity

$$\prod_{j=0}^{k} p_j^{\sum_{i \geq 1}(\chi_j(\varepsilon_i) - \chi_j(\varepsilon_i'))}$$

is a well-defined finite number, since all but a finite number of the terms in the infinite sums appearing in the exponent are zero. We denote this number by $J_{R^q(y)}$.

We now define the transformation $S : Y \longrightarrow Y$ as follows:

$$S(y) = R^q(y) \quad \text{if} \quad J_{R^t(y)} \neq 1 \quad \text{for} \quad 1 \leq t \leq q - 1 \quad \text{and} \quad J_{R^q(y)} = 1.$$

From the definition of the odometer transformation it is clear that when t goes through the set $\{0, 1, \ldots, (k + 1)^n - 1\}$, the first n-coordinates of the points $y, R(y), R^2(y), \ldots, R^{(k+1)^n - 1}(y)$ go over every element of the set $\{0, 1, \ldots, k\}^n$ exactly once. This implies that the transformation is well-defined on Y and is measurable with respect to $\mathscr{B}_Y$. Furthermore, by considering the action of R^{-1} it is easy to verify that S is bijective and S^{-1} is measurable. If we show that S is also nonsingular with respect to the product measure m then we can conclude that S is an element of the full group $[R]$.

We can actually show more. In fact S is not only nonsingular but actually preserves the measure m. To see this let us define $A_q = \{y \in Y : S(y) = R^q(y)\}$ for each $q \in \mathbb{N}$. Then $Y = \bigcup_{q>0} A_q (disj)$. Next we consider a cylinder set in $\mathscr{B}_Y$ of the form $E = \{y = (\varepsilon_1, \varepsilon_2, \ldots) \in Y : \varepsilon_{t_i} = a_i\}$ for $i = 1, 2, \ldots, s$, where $t_1 < t_2 < \cdots < t_s$ and for each i , a_i is a member of the set $\{0, 1, \ldots, k\}$. Since on the set $E \cap A_q$ the condition $J_{R^q(y)} = 1$ implies that R^q preserves the measure m, we get

$$m(SE) = m\left(S\left(\bigcup_{q>0} E \cap A_q\right)\right) = \sum_{q=1}^{\infty} m(S(E \cap A_q))$$

$$= \sum_{q=1}^{\infty} m(R^q(E \cap A_q)) = \sum_{q=1}^{\infty} m(E \cap A_q) = m(E).$$

As the cylinder sets generate the σ-algebra $\mathscr{B}_Y$, it follows that m is an invariant measure for S.

Next we note that for every element π of the permutation group $\mathscr{G}$, the transformation Ψ_π defined on $(Y, \mathscr{B}_Y, m)$ considered above is not only a member of the full group $[R]$, but is a member of the full group $[S]$, which is a subgroup of $[R]$. To see this, we note that if π belongs to $\mathscr{G}_n$, then the transformation Ψ_π permutes the first n coordinates of each point $y \in Y$ and leaves other coordinates unchanged. Therefore for each $u \in \mathbb{N}$ the quantity $J_{\Psi_\pi^u(y)}$, which is defined in the same way as for $J_{R^q(y)}$ given above, equals i for every y. Thus we have the following inclusion relationship:

$$\mathscr{G} \subset [S] \subset [R],$$

where we denote by the same symbol $\mathscr{G}$ the group of transformations $\{\Psi_\pi : \pi \in \mathscr{G}\}$. We use in the sequel the symbol $\mathscr{G}_n$ also to denote the transformation group $\{\Psi_\pi : \pi \in \mathscr{G}_n\}$. To show that the transformation S on $(Y, \mathscr{B}_Y, m)$ is ergodic it is enough to show that the action of the full group $[S]$ on $(Y, \mathscr{B}_Y, m)$ is ergodic in the sense that any set $B \in \mathscr{B}_Y$ invariant under every transformation in $[S]$ must satisfy $m(B) = 0$ or $m(B^c) = 0$. Because of the inclusion relation shown above, it is enough to show that the action of the subgroup $\mathscr{G}$ is ergodic. To prove this we use the following well-known fact from probability theory due to Hewitt and Savage:

The Hewitt–Savage Zero–One Law. *Any set $B \in \mathscr{B}_Y$ invariant under the action of the group $\mathscr{G}$ satisfies either $m(B) = 0$ or $m(B^c) = 0$.*

We give a short proof of this.

Proof. Let B be a set invariant under the action of the group $\mathscr{G}$. Then for any $\varepsilon > 0$, we can find a finite union E of cylinder sets, each of which depends only on at most the first n coordinates for some n, such that $m(B \triangle E) < \varepsilon$. Now take the permutation $\pi \in \mathscr{G}_{2n}$ defined by

$$\pi(i) = \begin{cases} i + n & \text{if } 1 \leq i \leq n, \\ i - n & \text{if } n + 1 \leq i \leq 2n. \end{cases}$$

Since for each cylinder set F that depends on at most the first n coordinates the set $\Psi_\pi(F)$ is a cylinder set that depends only on the $(n+1)$-th through $2n$-th coordinates, we have

$$m(F \cap \Psi_\pi(F)) = m(F)m(\Psi_\pi(F)) = m(F)^2.$$

Here we used the fact that m is a product measure invariant for the transformation Ψ_π. Therefore, we have $m(E \cap \Psi_\pi(E)) = (m(E))^2$.

Since by assumption the set B is invariant under the transformation Ψ_π, we have $B = B \cap \Psi_\pi(B)$. Since $0 < m(B)$, and $m(E) \leq m(X) = 1$, we get

$$0 \leq m(B) - m(B)^2 = m(B) - m(E \cap \Psi_\pi(E)) + (m(E))^2 - (m(B))^2$$

$$= [m(B \cap \Psi_\pi(B)) - m(B \cap \Psi_\pi(E))] + [m(B \cap \Psi_\pi(E))$$

$$- m(E \cap \Psi_\pi(E))] + (m(E))^2 - (m(B))^2$$

$$\leq |m(\Psi_\pi(B) \triangle \Psi_\pi(E))| + m(B \triangle E)(1 + m(B) + m(E))$$

$$\leq |m(B \triangle E)|(2 + m(B) + m(E))$$

$$\leq 4m(B \triangle E) < 4\varepsilon.$$

Since this holds for any $\varepsilon > 0$ we conclude that $m(B) = (m(B))^2$, which implies $m(B) = 0$ or $m(B) = 1$. □

From the Hewitt–Savage Zero–One Law it follows that the action of the group $\mathscr{G}$ on $(Y, \mathscr{B}_Y, m)$ is ergodic, which in turn implies that the transformation S is also ergodic. This argument likewise shows that the transformation R must be ergodic because of the inclusion relation shown above. But this last fact implies that unless the condition $p_0 = \cdots = p_k = \frac{1}{k+1}$ is satisfied the transformation R does not preserve any finite or σ-finite measure on $(Y, \mathscr{B}_Y)$ equivalent with m. For, if the condition is not satisfied, it is clear that m is not invariant under R, while if μ is a finite or σ-finite measure equivalent with m and invariant under R, then it is clear that μ is invariant under any element of the full group $[R]$ and in particular under the transformation S. But since S preserves m and is ergodic, the measure μ must be a constant multiple of the measure m, and this is impossible since R does not preserve m.

4.2.3   Construction of the Second Basic Example

We consider a special case of the above discussion; see also [31].

We construct an example of an infinite ergodic transformation T and a dissipative transformation Q defined on the σ-finite measure space $(X, \mathscr{B}, m)$ with the property that $TQ = QT$ and T preserves the measure m, while Q does not preserve the measure m. Among other things, by Corollary 5.1.4, this implies that every *eww* set for the ergodic transformation T has infinite measure.

Let $\alpha > 0$ be a positive number, and let $\beta = \alpha/(1 + \alpha)$. We consider the one-sided infinite direct product measure space

$$(Y, \mathscr{B}_Y, m) = \prod_{i=1}^{\infty}(Y_i, \mathscr{B}_i, m_i),$$

where $Y_i = \{0, 1\}$, $\mathscr{B}_i$ = all subsets of Y_i, with $m_i(0) = \beta$ and $m_i(1) = 1 - \beta$ for all $i \in \mathbb{N}$. We remove from Y the countable set

$$N = \{y = (\varepsilon_1, \varepsilon_2, \ldots) \in Y : \varepsilon_i = 0 \text{ for all } i \geq n \text{ and some } n \in \mathbb{N} \text{ or}$$

$$\varepsilon_i = 1 \text{ for all } i \geq n \text{ and some } n \in \mathbb{N}\}.$$

For each $j \in \mathbb{N}$ we consider the cylinder set

$$Y[j] = \{y = (\varepsilon_1, \varepsilon_2, \ldots) \in Y : y = (\underbrace{1, \ldots 1}_{j-1 \; times}, 0, \varepsilon_{j+1}, \ldots)\},$$

and define the transformation R by

$$R(Y[j]) = \{y = (\varepsilon_1, \varepsilon_2, \ldots) \in Y : y = (\underbrace{0, \ldots 0}_{j-1 \; times}, 1, \varepsilon_{j+1}, \ldots)\}.$$

It is clear that $Y = \bigcup_{j=1}^{\infty} Y[j]$. We also have for $j = 1, 2, \ldots$

$$m(Y[j]) = (1 - \beta)^{j-1}\beta, \text{ and } m(RY[j]) = \alpha^{j-2}m(Y[j]).$$

Figure 4.3 illustrates the action of R on Y.

We consider the measure space $(X, \mathscr{B}, m)$, where

$$X = Y \times \mathbb{Z} = \{(y, n) : y \in Y, n \in \mathbb{Z}\},$$

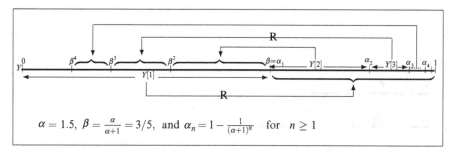

Fig. 4.3 The transformation R on the space Y

$\mathscr{B}$ is the σ-field generated by all sets of the form

$$(B,n) = \{(y,n) \in X : y \in B, \quad B \in \mathscr{B}_Y\} \quad \text{for} \quad n \in \mathbb{Z}$$

and on $(X, \mathscr{B})$, we write m for the extension of the measure m on $(Y, \mathscr{B}_Y)$; the extension of m to X is defined by:

$$m(B,n) = \alpha^n m(B) \quad \text{for} \quad (B,n) \subset (Y,n) \quad \text{and} \quad n \in \mathbb{Z}.$$

Next we define the transformation Q on $(X, \mathscr{B}, m)$ as follows:

$$Q(y,n) = (y, n+1).$$

It is easy to see that Q is a dissipative transformation defined on the measure space $(X, \mathscr{B}, m)$, where the set $(Y, 0)$ is a wandering set for the transformation Q. We have $X = \bigcup_{n \in \mathbb{Z}} Q^n(Y,0) \, (disj)$, and

$$m(QA) = \alpha m(A) \quad \text{for all} \quad A \in \mathscr{B}. \tag{4.24}$$

We define the transformation T on $(X, \mathscr{B}, m)$ as follows:
Let $x = (y,n) \in X$ where $y \in Y[j]$ for $j \in \mathbb{N}$ and $n \in \mathbb{Z}$. Then

$$T(x) = T(y,n) = (Ry, n + j - 2). \tag{4.25}$$

It is clear that T is a 1-1, measurable and nonsingular transformation defined on the infinite measure space $(X, \mathscr{B}, m)$, and $TQ = QT$. Figure 4.4 displays the action of T and Q on X.

It remains to show that T is an ergodic measure-preserving transformation on the infinite measure space $(X, \mathscr{B}, m)$.

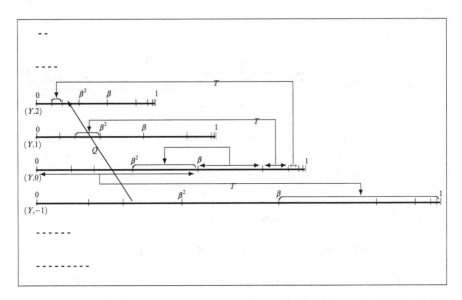

Fig. 4.4 The Second Basic Example T and the dissipative commutator Q acting on X

Let us consider a set $(A, n) \subset (Y[j], n)$ for $n \in \mathbb{Z}$ and $j \in \mathbb{N}$. From (4.25) follows:

$$m(T(A, n)) = m(RA, n + j - 2) = \alpha^{j-2} m(A, n + j - 2) = m(A, n).$$

The fact that any measurable set $B \subset X$ is a countable union of sets of the above form implies that m is an invariant measure for T.

To show that T is an ergodic transformation we consider the following:
For a point $y = (\varepsilon_1, \varepsilon_2, \ldots) \in Y$ and an integer $k \in \mathbb{N}$ we define the k-segment of y to be $I_k(y) = (\varepsilon_1, \ldots, \varepsilon_k)$. We also let $J(y) = \{n \in \mathbb{N} : I_{n+2}(y) = (\varepsilon_1, \ldots, \varepsilon_n, 1, 0)\}$. It is clear that for all $y \in Y$,

$$J(y) \text{ contains infinitely many integers } n \in \mathbb{N}. \tag{4.26}$$

Let us call a k-tuple $(\eta_1, \ldots, \eta_k)$ a k-block if $\eta_i = 0$ or 1 for $i = 1, 2, \ldots, k$.

It is not difficult to show that for each $y = (\varepsilon_1, \varepsilon_2, \ldots) \in Y$ and $k \in J(y)$ the transformation R defined on $(Y, \mathscr{B}_Y, m)$ satisfies the following:

$$\text{each of the } 2^k \text{ distinct } k\text{-blocks } (\eta_1, \ldots, \eta_k) \text{ will appear as a}$$

$$k\text{-segment } I_k(R^j y) \text{ for some } j \text{ with } 1 \leq j \leq 2^k. \tag{4.27}$$

We also have: for $y \in Y$ and $k \in J(y)$, if $I_k(R^j y) = (\eta_1, \ldots, \eta_k)$ for some j with $1 \leq j \leq 2^k$, then

$$T^j(y, n) = (R^j y, n - r(k, j)) \text{ where } r(k, j) = \sum_{i=1}^{k} (\varepsilon_i - \eta_i).$$

For a point $y \in Y$ and integers $n, k \in \mathbb{Z}$ using (4.26) we choose a positive integer $q = q(y, n, k) \in J(y)$ such that $\sum_{i=1}^{q} \varepsilon_i \geq n - k \geq \sum_{i=1}^{q} (\varepsilon_i - 1)$.

Next, using 4.27 we choose an integer p satisfying $1 \leq p \leq 2^q$ such that if $R^p y = (\eta_1, \eta_2, \ldots)$ then $\sum_{i=1}^{q} (\varepsilon_i - \eta_i) = n - k$.

For this choice of p we have $T^p x \in (Y, k)$. Therefore we conclude:

For all $x = (y, n) \in X$ and $k \in \mathbb{Z}$, $\exists \ p > 0$ such that $T^p x \in (Y, k)$.

Using the above, we define the transformation S induced by T on the subset $(Y, 0)$ as follows:

For $(y, 0) \in (Y, 0)$, let $S(y, 0) = T^s(y, 0)$, where $s = \min\{p \in \mathbb{N} : T^p(y, 0) \in (Y, 0)\}$.

Using the Hewitt–Savage Zero–One Law we conclude that $S : (Y, 0) \longrightarrow (Y, 0)$ is an ergodic transformation, which implies that T is ergodic. We also have that the transformation Q, which is not measure-preserving, is in the centralizer of T.

By identifying Y with the subset $(Y, 0)$ in X, we will "extend" the nonsingular transformation R on Y to a measure-preserving transformation T on X. The transformation S induced on the subset $(Y, 0)$ by T is the transformation used by Kakutani in [45], and also the transformation that Vershik calls the Pascal automorphism [54].

Just like the First Basic Example the transformation T of the Second Basic Example is also an infinite ergodic transformation that possesses recurrent sequences. The following proposition exhibits a collection of recurrent sequences for it.

Proposition 4.2.3. *For any integer $k > 0$ the sequence $\{2^n k : n = 0, 1, 2, \ldots\}$ is a recurrent sequence for the transformation T of the Second Basic Example.*

Proof. Fix an integer $k > 0$ and let $\eta_0 \ldots \eta_{i-1} 0$ be the base-2 representation of $2k$ with the highest bit ($\eta_0 = 1$) on the left. Consider the cylinder A_0 in Y based on fixing the first $i + 1$ coordinates as $et a_0, \eta_1, \ldots, \eta_{i-1}, 0$: we denote this cylinder by

$$A_0 = (\eta_0, \eta_1, \ldots, \eta_{i-1}, 0, \varepsilon_{i+2}, \ldots)$$

For $n \geq 1$ let A_n be the n-shifted cylinder A_0

$$A_n = (\varepsilon_1, \ldots, \varepsilon_n, \eta_0, \eta_1, \ldots, \eta_{i-1}, 0, \varepsilon_{n+i+2}, \ldots)$$

where the ε_k's are either 0 or 1.

Then

$$R^{2^n k} A_n = (\varepsilon_1, \ldots, \varepsilon_n, 0, \eta_0, \eta_1, \ldots, \eta_{i-1}, \varepsilon_{n+i+2}, \ldots).$$

It is clear that for any $n \geq 0$

$$T^{2^n k}(A_n, 0) = (R^{2^n k} A_n, 0) \subset (Y, 0),$$

and

$$m\big(T^{2^n k}(A_n, 0)\big) = m(A_0).$$

This implies

$$m\big(T^{2^n k}(Y, 0) \cap (Y, 0)\big) \geq m(R^{2^n k} A_n, 0) = m(A_0, 0) > 0 \quad \text{for all } n \geq 0.$$

Proposition 3.3.2 then says: for any integer $k > 0$, $\{2^n k : n = 0, 1, 2, \ldots\}$ is a recurrent sequence for the transformation T.   □

4.3   Third Basic Example

In the previous two sections we discussed examples of infinite ergodic transformations with several interesting properties which also possessed recurrent sequences. In this section we present two infinite ergodic transformations that do not possess recurrent sequences. These transformations are important for Sect. 3.3.2.

4.3.1   Construction of the Third Basic Example

As in the First Basic Example above, let us consider the measure space $(X_0, \mathcal{B}_0, m_0)$, for $n \geq 0$ the intervals $A_n = \{x \in X_0 : 1 - \frac{1}{2^n} < x < 1 - \frac{1}{2^{n+1}}\}$ and the transformation S on X_0 defined by: $Sx = x - \left(1 - 3/2^{n+1}\right)$ for $x \in A_n$, $n \geq 0$.

For $n \geq 0$ we divide each interval A_n into 2^n equal subintervals; namely

$$A_{n,i} = \{x \in X_0 : 1 - \frac{1}{2^n} + \frac{i}{2^{2n+1}} < x < 1 - \frac{1}{2^n} + \frac{i+1}{2^{2n+1}}\} \text{ for } i = 0, 1, \ldots, 2^n - 1.$$

Then

$$A_n = \bigcup_{i=0}^{2^n - 1} A_{n,i} \ (disj).$$

We note that $m(A_n) = \dfrac{1}{2^{n+1}}$ for $n \geq 0$, and

$$m(A_{n,i}) = \frac{1}{2^n 2^{n+1}} \quad \text{for } n \geq 0, \ 0 \leq i < 2^n - 1. \tag{4.28}$$

Let us put $f(0) = 0$, $f(1) = 2f(0) + 2$ and in general
$f(p) = 2f(p-1) + 2$ $(= 2^{p+1} - 2)$ for $n \geq 0$, $i = 0, 1, \ldots, 2^i - 1$.

Next, for each $n \geq 0$ and $i = 0, 1, 2, \ldots, 2^n - 1$ let us consider the sets $A_{n,i}$
and the integers $p = p[n, i] = 2^n - 1 + i$.

Again, as in the First Basic Example, let us consider $f(p) + 1$ mutually disjoint
copies of $A_{n,i}$; namely $A_{n,i}^0 = A_{n,i}$, $A_{n,i}^1$, $\ldots$, $A_{n,i}^{f(p)}$, and denote by the same letter
R all the isomorphisms:

$$R(A_{n,i}^k) = A_{n,i}^{k+1}, \quad \text{for } k = 0, 1, \ldots, f(p[n, i]) - 1, \ n \geq 0 \text{ and } i = 0, 1, 2, \ldots, 2^n - 1.$$

The measure space $(X, \mathscr{B}, m)$ is defined similarly, where

$$X = \bigcup_{n=0}^{\infty} \bigcup_{i=0}^{2^n-1} \bigcup_{k=0}^{f(p[n,i])} A_{n,i}^k (disj).$$

Then the infinite ergodic transformation T on $(X, \mathscr{B}, m)$ is defined by:

$$Tx = \begin{cases} Rx & \text{if } x \in A_{n,i}^k \text{ for } k = 0, 1, \ldots, f(p[n, i]) - 1 \text{ for } n \geq 0, \ 0 \leq i < 2^n, \\ S(R^{-f(p[n,i])}x) & \text{if } x \in A_{n,i}^{f(p[n,i])}, \ n \geq 0, \ 0 \leq i < 2^n. \end{cases}$$

We observe the following:
For $n \geq 0$, $i = 0, 1, \ldots, 2^n - 1$, $p = p[n, i] = 2^n - 1 + i$ and $f(p) = 2^{p+1} - 2$

$$T^{f(p[n,i])+2} A_0 \subset \bigcup_{r=n+1}^{\infty} \bigcup_{i=0}^{2^r-1} \bigcup_{k=0}^{f(p[r,i])} T^k A_{r,i}.$$

Also $T^{f(p[n,i])} A_0 \cap \displaystyle\bigcup_{k=0}^{f(p[n,i])} T^k A_{n,i}$ covers at most 2^{n-1} levels of $\displaystyle\bigcup_{k=0}^{f(p[n,i])} T^k A_{n,i}$.

This implies $m(T^n A_0 \cap A_0) \leq 2^{n-1} m(A_{n,i}) = \dfrac{2^{n-1}}{2^n 2^{n+1}} \longrightarrow 0$ as $n \to \infty$.

According to Lemma 3.3.8 and Remark 3.3.3 it follows that the transformation
T is an infinite ergodic transformation without recurrent sequences.

Figure 4.5 describes the transformation T.

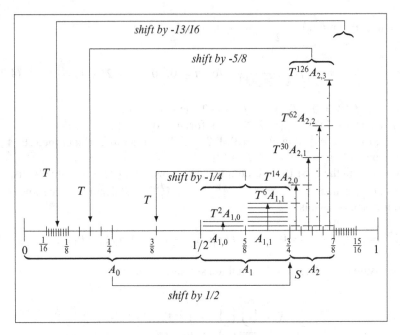

Fig. 4.5 The Third Basic Example T on the space X

4.3.2 Random Walk on the Integers

We sketch the next example, which is essentially a realization of simple random walk on the integers as an infinite ergodic transformation $\tilde{T}$ on the two-sided infinite strip. This is also an infinite ergodic transformation with no recurrent sequences. A variation of this example was discussed by E. Hopf in [38].

On the unit square I^2 let us consider the Baker's map $T : I^2 \to I^2$,

$$
T(z, y) = \begin{cases} (2z, y/2) & \text{if } 0 \le z < \frac{1}{2}, \\ (2z, y/2 + 1/2) & \text{if } \frac{1}{2} \le z < 1, \end{cases}
$$

and the function $\phi : I^2 \to \{-1, 1\}$ defined by

$$
\phi(z, y) = \begin{cases} -1 & \text{if } 0 \le z < \frac{1}{2}, \\ +1 & \text{if } \frac{1}{2} \le z < 1. \end{cases}
$$

The transformation $\tilde{T}$ is the skew product on $X = \mathbb{Z} \times I^2$ defined by

$$
\tilde{T}(i, (z, y)) = (i + \phi(z, y), T(z, y)).
$$

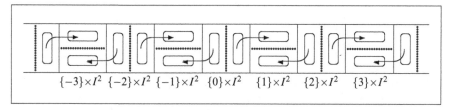

$\{-3\}\times I^2 \;\; \{-2\}\times I^2 \;\; \{-1\}\times I^2 \;\; \{0\}\times I^2 \qquad \{1\}\times I^2 \qquad \{2\}\times I^2 \qquad \{3\}\times I^2$

Fig. 4.6 The action of the $\tilde{T}$ on the even-numbered boxes of the space $X = \mathbb{Z} \times I^2$: $\tilde{T}$ is a realization of random walk on $\mathbb{Z}$

$\tilde{T}$ is an infinite measure-preserving transformation with the measure μ on X being the product of counting measure on $\mathbb{Z}$ and Lebesgue measure on the square. Thus the space X, which is the union of countably many copies of the square, is just the strip $\mathbb{R} \times [0, 1)$ with Lebesgue measure. Figure 4.6 gives a diagram of the action of $\tilde{T}$ on the even-numbered boxes in the infinite strip. Since

$$\tilde{T}^n(i, (z, y)) = (i + S_n(\phi(z, y)), T^n(z, y)),$$

where $S_n(\phi(z, y)) = \phi(z, y) + \phi(T(z, y)) + \cdots + \phi(T^{n-1}(z, y))$ is simple random walk on the integers; in other words, S_n is the sum of n independent Bernoulli random variables.

Using standard arguments, it is possible to show that $\tilde{T}$ is an infinite ergodic transformation on the space $X = \mathbb{Z} \times I^2$. By making use of Stirling's formula it is possible to show that $\tilde{T}$ is an infinite ergodic transformation without recurrent sequences. Namely, for $A_0 = \{0\} \times I^2$, we have $\lim_{n \to \infty} \mu\big(\tilde{T}^n A_0 \cap A_0\big) = 0$.

Chapter 5
Properties of Various Sequences

In this chapter we discuss properties of infinite sequences of integers associated with an infinite ergodic transformation. We note that all the sequences we consider are isomorphism invariants for such transformations.

We establish the following notation: $\mathbb{Z}$ is the set of all integers, and $\mathbb{N}$ is the set of all nonnegative integers. For subsets $\mathbb{A}$, $\mathbb{B} \subset \mathbb{Z}$ and an integer $k \in \mathbb{Z}$ let

$\mathbb{A} + \mathbb{B} = \{a + b : a \in \mathbb{A},\ b \in \mathbb{B}\}$, the *sumset* of $\mathbb{A}$ and $\mathbb{B}$,

$\mathbb{A} - \mathbb{B} = \{a - b : a \in \mathbb{A},\ b \in \mathbb{B}\}$, the *difference set* of $\mathbb{A}$ and $\mathbb{B}$,

$k\mathbb{A} = \{ka : a \in \mathbb{A}\}$,

$-\mathbb{A} = \{-a : a \in \mathbb{A}\}$,

$\mathbb{A}^c = \{n \in \mathbb{Z} : n \notin \mathbb{A}\}$,

$\mathbb{A} \smallsetminus \mathbb{B} = \mathbb{A} \cap \mathbb{B}^c$,

$|\mathbb{A}| =$ the number of elements in $\mathbb{A}$.

5.1 Properties of *ww* and Recurrent Sequences

For an infinite ergodic transformation T, let us consider the set $\mathscr{W} = \{\mathbb{W}\}$ of all *ww* sequences for T. It is clear that $\mathscr{W}$ is an isomorphism invariant; see [27].

Proposition 5.1.1. *Let T be a measure-preserving ergodic transformation defined on the infinite measure space $(X, \mathscr{B}, m)$, and let $\mathscr{W} = \{\mathbb{W}\}$ be the collection of all ww sequences $\mathbb{W} = \{w_i : i = 1, 2, \cdots\}$ for the transformation T. Then*

(i) $\mathbb{W} \in \mathscr{W} \implies -\mathbb{W} \in \mathscr{W}$.

(ii) $\mathbb{W} \in \mathscr{W}$, $\mathbb{V}$ *an infinite subsequence of* $\mathbb{W} \implies \mathbb{V} \in \mathscr{W}$.

(iii) $\mathbb{W} \in \mathscr{W} \implies \mathbb{W} + \{k\} \in \mathscr{W}$ *for any* $k \in \mathbb{Z}$.

(iv) $\mathbb{W} \in \mathscr{W} \implies \exists$ *an integer* $k > 0$ *such that*
$$\mathbb{W} \cap (\mathbb{W} + \{k\}) = \emptyset \quad \text{and} \quad \mathbb{W} \cup (\mathbb{W} + \{k\}) \in \mathscr{W}.$$

© Springer Japan 2014
S. Eigen et al., *Weakly Wandering Sequences in Ergodic Theory*,
Springer Monographs in Mathematics, DOI 10.1007/978-4-431-55108-9_5

(v) $\mathbb{W} \in \mathcal{W} \implies$ *for any* $k > 0$, $(\mathbb{W} - \mathbb{W})^c \cap k\mathbb{N}$ *must be an infinite subset of* $k\mathbb{N}$.

(vi) $\mathbb{W} = \{w_i : i = 1, 2, \dots\} \in \mathcal{W} \implies \lim\limits_{i \to \infty} \frac{i}{w_i} = 0$.

Proof. (i), (ii), and (iii) follow from the definitions.

To prove (iv) we let A be a *ww* set with the sequence $\mathbb{W} = \{w_i\}$ and note that $m(A) > 0$. Since T is ergodic there exists an integer $k > 0$ such that $m(T^{-k} A \cap A) > 0$. Let $B = T^{-k} A \cap A$. By possibly considering a subset of B we may assume $0 < m(B) < \infty$. Let $C = B \smallsetminus T^k B$. Since T is ergodic and $m(X) = \infty$ it follows that $m(C) > 0$.

Since $C \subset B \subset A$, we have $T^{w_i} C \cap T^{w_j} C = \emptyset$ for $i \neq j$.

Furthermore,

$$C \subset A \text{ and } T^k C \subset A \implies T^{w_i} C \cap T^{w_j + k} C = \emptyset \quad \text{for} \quad i \neq j,$$

and

$$C \subset B, \ T^k C \subset T^k B \text{ and } C \cap T^k B = \emptyset \implies C \cap T^k C = \emptyset.$$

Thus we arrive at the following: for some $k > 0$, $\mathbb{W} \cap (\mathbb{W} + \{k\}) = \emptyset$, and

$$T^{w_i'} C \cap T^{w_j'} C = \emptyset \text{ for } w_i' \neq w_j'; \ w_i', w_j' \in \mathbb{W} \cup (\mathbb{W} + \{k\}), \text{ and } i, j = 1, 2, \dots.$$

This shows that for some $k > 0$, $\mathbb{W} \cap (\mathbb{W} + \{k\}) = \emptyset$ and $\mathbb{W} \cup (\mathbb{W} + \{k\}) \in \mathcal{W}$.

To show (v) we note that if A is a *ww* set with the sequence $\mathbb{W} = \{w_i\}$ and $p \in \mathbb{W} - \mathbb{W}$ with $p \neq 0$, then $T^p A \cap A = \emptyset$.

If for some $k > 0$, statement (v) is not true, then the set $(\mathbb{W} - \mathbb{W}) \cap k\mathbb{N}$ must contain all but a finite number of elements of $k\mathbb{N}$. This implies there exists an integer $q > 0$ such that

$$T^{kn} A \cap A = \emptyset \quad \text{for} \quad n = q, q + 1, \dots,$$

and from this follows that A is a wandering set of positive measure for the transformation T^k. This is impossible since T is ergodic.

Finally, statement (vi) is a consequence of the following. Let A be any set of finite positive measure, which is *ww* for $\{w_i : i \geq 1\} = \mathbb{W} \in \mathcal{W}$. Without loss of generality we can assume $\lim w_i = \infty$ (by replacing $\mathbb{W}$ by $-\mathbb{W}$ if necessary). Since T is infinite ergodic, Lemma 5.2.6 implies

$$\lim_{i \to \infty} \frac{1}{w_i} m \left(\bigcup_{k=1}^{w_i} T^k A \right) = 0.$$

From this we conclude

$$0 = \lim_{i \to \infty} \frac{1}{w_i} m\left(\bigcup_{k=1}^{i} T^{w_k} A \right) = \lim_{i \to \infty} \frac{i}{w_i} m(A),$$

as required for (vi).   □

Initially, when it was noticed that an infinite ergodic transformation possesses *ww* sets, for a brief moment it was hoped that such sets might always be of finite measure. However, it was shown in [25] that this was no more than wishful thinking.

Next we consider *eww* sets and sequences for an infinite ergodic transformation.

As we saw in Chap. 2, an infinite ergodic transformation defined on a σ-finite measure space $(X, \mathscr{B}, m)$ does not preserve a finite invariant measure $\mu \sim m$. Therefore, Theorem 2.2.3 holds for such transformations, and we may conclude that every infinite ergodic transformation not only possesses *ww* sequences but it possesses a special kind of *eww* sequence as well; namely, what we call a *hereditary eww* sequence. By this we mean an infinite sequence of integers with the property that every infinite subsequence of it is again an *eww* sequence. Moreover, it is not difficult to show that for an infinite ergodic transformation all the *eww* sets for the hereditary *eww* sequences constructed by the methods of Chap. 2 have infinite measure. The infinite ergodic transformation of the Second Basic Example of Chap. 4 has the property that all of its *eww* sets are of infinite measure. However, in Chap. 4 it was pointed out that the infinite ergodic transformation of the First Basic Example admitted an *eww* set of finite measure. This was a surprising and interesting fact. In the next theorem we show the significance of this by exhibiting some interesting properties of an infinite ergodic transformation that possesses an *eww* sequence $\{n_i\}$ which has an *eww* set W of finite measure (see [14]).

Let us say that a set A is an *exhaustive* (*exh*) set for a transformation T if for some sequence $\{n_i : i \geq 1\}$, $X = \bigcup_{i=1}^{\infty} T^{n_i} A$. In this case we shall say that $\{n_i : i \geq 1\}$ is an *exh* sequence for T. We note that the sequence of sets $\{T^{n_i} A\}$ are not necessarily mutually disjoint.

Theorem 5.1.2. *Let T be an infinite ergodic transformation defined on the infinite measure space $(X, \mathscr{B}, m)$. Suppose T has an eww set W of finite measure with eww sequence $\{n_i : i \geq 1\}$. For a set $V \in \mathscr{B}$ let us consider the following statements:*

(i) V is an exh set with the sequence $\{n_i\}$.
(ii) V is a ww set with the sequence $\{n_i\}$.
(iii) $m(V) = m(W)$.

Then any two of the above statements together imply the third. For the implication (i) and (ii) together implying (iii), the condition $m(W) < \infty$ is not necessary.

By virtue of this theorem we define an infinite ergodic T to be of *finite type* if T admits an *eww* set of finite measure. Before proving Theorem 5.1.2 we state and prove a lemma where we gather some properties of sets that are *exh* or *ww*.

Lemma 5.1.3. *For an infinite ergodic transformation T the following statements are true:*

(i) W ww with the sequence $\{n_i\}$ $\Longrightarrow$ W ww with the sequence $\{-n_i\}$.

(ii) W ww with the sequence $\{n_i\}$ $\Longrightarrow$ $T^k W$ ww with the sequence $\{n_i\}$ for all $k \in \mathbb{Z}$.

(iii) W exh with the sequence $\{n_i\}$ $\Longrightarrow$ $T^k W$ exh with the sequence $\{n_i\}$ for all $k \in \mathbb{Z}$.

(iv) W ww with the sequence $\{n_i\}$ and V exh with the sequence $\{n_i\}$ $\Longrightarrow$ $m(W) \le m(V)$.

(v) W eww with the sequence $\{n_i\}$ and $m(W) < \infty$ $\Longrightarrow$ W is eww with the sequence $\{-n_i\}$.

Proof. (i), (ii), and (iii) follow from the definitions.

We prove (iv). Since V is *exh* with the sequence $\{n_i\}$, T is measure-preserving, and W is *ww* with the sequence $\{-n_i\}$ we have:

$$m(W) = m(W \cap \bigcup_{i=1}^{\infty} T^{n_i} V) \le \sum_{i=1}^{\infty} m(W \cap T^{n_i} V) = \sum_{i=1}^{\infty} m(T^{-n_i} W \cap V)$$

$$= m(\bigcup_{i=1}^{\infty} T^{-n_i} W \cap V) \le m(V).$$

To prove (v) it is enough to show that W is *exh* with the sequence $\{-n_i\}$. For each $k \in \mathbb{Z}$ we let $W_k = T^k W$. Since W_k is *eww* with the sequence $\{n_i\}$, T is measure-preserving, and W is *ww* with the sequence $\{-n_i\}$, we have:

$$m(W) = m(W \cap \bigcup_{i=1}^{\infty} T^{n_i} W_k) \le \sum_{i=1}^{\infty} m(W \cap T^{n_i} W_k) = \sum_{i=1}^{\infty} m(T^{-n_i} W \cap W_k)$$

$$= m(\bigcup_{i=1}^{\infty} T^{-n_i} W \cap W_k) \le m(W_k) = m(W) < \infty.$$

This says that for each $k \in \mathbb{Z}$, $W_k = T^k W \subset \bigcup_{i=1}^{\infty} T^{-n_i} W$. Since T is ergodic we have $\bigcup_{k=1}^{\infty} T^k W = X$ and conclude $X = \bigcup_{i=1}^{\infty} T^{-n_i} W$. □

Proof (Theorem 5.1.2). We suppose (i) holds together with (ii). We do not need to assume $m(W) < \infty$. From (iv) of Lemma 5.1.3 follows that $m(W) \le m(V)$. Interchanging the roles of V and W, we conclude (iii).

For the remainder of the proof we assume $m(W) < \infty$. For each $k \in \mathbb{Z}$ we let $W_k = T^k W$. Since W_k is an *eww* set with the sequence $\{-n_i\}$ and T is measure-preserving, using Lemma 5.1.3 we get

$$m(V) = m\left(\bigcup_{i=1}^{\infty} T^{-n_i} W_k \cap V\right) = \sum_{i=1}^{\infty} m(T^{-n_i} W_k \cap V)$$

$$= \sum_{i=1}^{\infty} m(W_k \cap T^{n_i} V). \tag{5.1}$$

Now suppose (ii) holds, together with (iii). Since V is *ww* with the sequence $\{n_i\}$ and T is measure-preserving, using (5.1) we get

$$m(V) = \sum_{i=1}^{\infty} m(W_k \cap T^{n_i} V) = m\left(W_k \cap \bigcup_{i=1}^{\infty} T^{n_i} V\right) \le m(W_k) = m(W).$$

This says that $T^k W \subset \bigcup_{i=1}^{\infty} T^{n_i} V$ for each $k \in \mathbb{Z}$. Since T is ergodic, we conclude that V is *exh* with the sequence $\{n_i\}$, and therefore (i) holds.

Finally, suppose (i) holds, together with (iii). Since V is *exh* with the sequence $\{n_i\}$ and T is measure-preserving, using (5.1) we get

$$m(V) = \sum_{i=1}^{\infty} (W_k \cap T^{n_i} V) \ge m\left(W_k \cap \bigcup_{i=1}^{\infty} T^{n_i} V\right) = m(W_k) = m(W).$$

This says that for each $k \in \mathbb{Z}$ and for $i \ne j$ the sets $T^{n_i} V$ and $T^{n_j} V$ do not intersect on the set W_k. Since T is ergodic we conclude that V is *ww* with the sequence $\{n_i\}$, and therefore (ii) holds. $\square$

Corollary 5.1.4. *Let T be an infinite ergodic transformation that possesses an eww set of finite measure. Then T admits only measure-preserving commutators. In other words, if Q is a measurable transformation satisfying $QT = TQ$ then Q preserves the measure m.*

Proof. For any set $A \in \mathcal{B}$ we let $m'(A) = m(QA)$; then $m' \sim m$.
Since $m'(TA) = m(QTA) = m(TQA) = m'(A)$, this says that m' is an invariant measure for T. Because T is ergodic, it follows that there exists a constant $\alpha > 0$ satisfying $m(QA) = \alpha m(A)$ for all $A \in \mathcal{B}$. If W is an *eww* set of finite measure for T with a sequence $\{n_i\}$ then so is the set QW with the same sequence $\{n_i\}$. Hence by Theorem 5.1.2, $m(QW) = m(W)$, which implies $\alpha = 1$. This says Q preserves the measure m. $\square$

In Chap. 3 we introduced the notion of a recurrent sequence for infinite ergodic transformations (Definition 3.3.1). We note that the collection of all recurrent sequences of a transformation is an isomorphism invariant for infinite ergodic transformations. In Proposition 3.3.2, we gave the following characterization of a recurrent sequence:

A sequence $\mathbb{R} = \{r_i : i = 1, 2, \ldots\}$ of integers is a recurrent sequence for T if there exists a set A of finite measure for which $\liminf_{i \to \infty} m(T^{r_i} A \cap A) > 0$.

We now list properties of the collection $\mathscr{R} = \{\mathbb{R}\}$ of all recurrent sequences $\mathbb{R}$ for an infinite ergodic transformation T (see [27]).

Proposition 5.1.5. *Let T be a measure-preserving ergodic transformation defined on the infinite measure space $(X, \mathscr{B}, m)$, and let $\mathscr{R} = \{\mathbb{R}\}$ be the collection of all recurrent sequences $\mathbb{R} = \{r_i : i = 1, 2, \ldots\}$ for the transformation T. Then*

(i) $\mathbb{R} \in \mathscr{R} \implies -\mathbb{R} \in \mathscr{R}$.
(ii) $\mathbb{R} \in \mathscr{R}$, $\mathbb{S} \subset \mathbb{R}$ ($\mathbb{S}$ an infinite subsequence of $\mathbb{R}$) $\implies \mathbb{S} \in \mathscr{R}$.
(iii) $\mathbb{R} \in \mathscr{R} \implies \mathbb{R} + \{k\} \in \mathscr{R}$ for any $k \in \mathbb{Z}$.
(iv) $\mathbb{R} \in \mathscr{R}$, $\mathbb{S} \in \mathscr{R} \implies \mathbb{R} \cup \mathbb{S} \in \mathscr{R}$.
(v) $\mathbb{R} = \{r_i\} \in \mathscr{R} \implies \lim_{i \to \infty} \frac{i}{r_i} = 0$.

Proof. (i), (ii), (iii), and (iv) follow from the definitions and the properties of the collection $\mathscr{W}$ of the *ww* sequences established in Proposition 5.1.1.

To prove (v), let A be a set of finite measure for which $\liminf\limits_{i \to \infty} m(T^{r_i} A \cap A) > 0$.

By possibly ignoring a finite number of terms we may assume that there exists a positive number α such that $m(T^{r_i} A \cap A) \geq \alpha$ for all $i \geq 1$. Then

$$\frac{1}{r_k} \sum_{j=1}^{r_k} m(T^j A \cap A) \geq \frac{1}{r_k} \sum_{j=1}^{k} m(T^{r_j} A \cap A) \geq \frac{1}{r_k}(k\alpha) = \frac{k}{r_k}\alpha .$$

Since $\lim\limits_{k \to \infty} \frac{1}{r_k} \sum\limits_{j=1}^{r_k} m(T^j A \cap A) = 0$ the above implies $\lim\limits_{k \to \infty} \frac{k}{r_k} = 0$. $\square$

5.2 Dissipative Sequences

We saw earlier that an ergodic transformation T is recurrent. That is, almost all points $x \in X$ visit every set A of positive measure infinitely often under images of T; in other words, the cardinality of the intersection of every set A of positive measure with $Orb_{(T,\mathbb{Z})}(x) = \{T^n x : n \in \mathbb{Z}\}$ is infinite for a.a. $x \in X$. In contrast to that, let us consider the following property for a sequence of integers $\mathbb{D} = \{d_i : i = 1, 2, \ldots\}$ with respect to an infinite ergodic transformation T defined on $(X, \mathscr{B}, m)$:

> For every set A of finite measure and for almost all points $x \in X$, $T^{d_i} x \in A$ occurs only for finitely many i's; in other words, the cardinality of the intersection of every set A of finite measure with $Orb_{(T,\mathbb{D})}(x) = \{T^n x : n \in \mathbb{D}\}$ is finite for a.a. $x \in X$.

More generally, we introduce the following definition:

Definition 5.2.1. A sequence $\mathbb{D} = \{d_i : i = 1, 2, \ldots\}$ of integers is called a *dissipative sequence* for a transformation T defined on an infinite measure space $(X, \mathscr{B}, m)$ if for every $f \in L^1(m)$,

$$\sum_{i=0}^{\infty} |f(T^{d_i}(x))| < \infty \quad \text{a.e. on } X. \tag{5.2}$$

If a set A has finite measure, then its characteristic function χ_A belongs to $L^1(m)$. Therefore, it is clear that a dissipative sequence for a transformation T satisfies the property mentioned above.

The next proposition and its corollary show that every infinite ergodic transformation has dissipative sequences.

Proposition 5.2.2. *Let T be an infinite ergodic transformation defined on $(X, \mathcal{B}, m)$. An infinite sequence $\mathbb{D} = \{d_i : i = 1, 2, \ldots\}$ of integers is a dissipative sequence for T if there exist a function $g \in L^\infty(m)$ with $g(x) > 0$ a.e., and a positive constant c for which*

$$\sum_{i=1}^{\infty} g(T^{-d_i} x) \le c \quad \textit{a.e. on } X. \tag{5.3}$$

Proof. Suppose for an infinite sequence $\mathbb{D} = \{d_i : i = 1, 2, \ldots\}$ there exist a function $g \in L^\infty(m)$ with $g(x) > 0$ and a positive constant c for which (5.3) is satisfied. Then, for any function $f \in L^1(m)$, we have

$$\infty > \int_X c|f(x)|dm(x) \ge \int_X \sum_{i=1}^{\infty} g(T^{-d_i}(x))|f(x)|dm(x)$$

$$= \int_X g(x) \sum_{i=1}^{\infty} |f(T^{d_i}(x)|dm(x).$$

From this follows that $g(x) \sum_{i=1}^{\infty} |f(T^{d_i}(x))| < \infty$ a.e. Since $g(x) > 0$, we conclude that the condition (5.2) is satisfied for every $f \in L^1(m)$, and therefore $\mathbb{D}$ is a dissipative sequence for T. $\qquad\square$

Corollary 5.2.3. *Every ww sequence $\mathbb{W}$ for an infinite ergodic transformation T is a dissipative sequence for T.*

Proof. Let A be a set of positive measure which is ww with the sequence $\mathbb{W} = \{w_i\}$. Since T is ergodic we have $\bigcup_{k=1}^{\infty} T^k A = X$. Let us define the function g by

$$g(x) = \sum_{k=1}^{\infty} \frac{1}{2^k} \chi_{T^k A}(x).$$

Then $0 < g(x) \le 1$ a.e. on X, and since for each $k \ge 1$ the set $T^k A$ is ww with the same sequence $\mathbb{W} = \{w_i\}$, $\sum_{i=1}^{\infty} \chi_{T^k A}(T^{-w_i}(x)) \le 1$ holds a.e. Therefore

$$\sum_{i=1}^{\infty} g(T^{-w_i}(x)) = \sum_{k=1}^{\infty} \frac{1}{2^k} \left(\sum_{i=1}^{\infty} \chi_{T^k A}(T^{-w_i}(x)) \right) \le \sum_{k=1}^{\infty} \frac{1}{2^k} = 1,$$

which shows that the function g satisfies (5.3) with $c = 1$. This says $\mathbb{W}$ is a dissipative sequence for T.   □

Proposition 5.2.4. *Let T be an infinite ergodic transformation defined on the infinite measure space $(X, \mathscr{B}, m)$. If $\mathbb{D} = \{d_i\}$ is a dissipative sequence for T, then*

$$m(A) < \infty, \ m(B) < \infty \implies \lim_{i \to \infty} m(T^{d_i} A \cap B) = 0. \qquad (5.4)$$

Proof. Let $\mathbb{D} = \{d_i\}$ be a dissipative sequence. If $m(B) < \infty$ then its characteristic function χ_B belongs to $L^1(m)$. Therefore, $\sum_{i=1}^{\infty} \chi_B(T^{d_i} x) < \infty$ a.e. This implies $\lim_{i \to \infty} \chi_B(T^{d_i} x) = 0$ a.e. Since $m(A) < \infty$ and for each i, $\chi_A(T^{d_i} x) \leq 1$ a.e., the dominated convergence theorem applies. Therefore

$$\lim_{i \to \infty} \int_A \chi_B(T^{d_i} x) dm(x) = \lim_{i \to \infty} m(A \cap T^{-d_i} B) = 0.$$

Since T is measure-preserving, $m(A \cap T^{-d_i} B) = m(T^{d_i} A \cap B)$ for each i, and we conclude $\lim_{i \to \infty} m(T^{d_i} A \cap B) = 0$.   □

Next, as in [27], we consider the collection $\mathscr{D}$ of all the dissipative sequences for an infinite ergodic transformation T. From Definition 5.2.1, it is clear that $\mathscr{D}$ is closed under the operation of taking finite unions. Corollary 5.2.3 implies that $\mathscr{D}$ contains $\mathscr{W}$.

We list properties of the collection $\mathscr{D}$ for the transformation T.

Proposition 5.2.5. *Let T be an infinite ergodic transformation defined on the infinite measure space $(X, \mathscr{B}, m)$, and let $\mathscr{D} = \{\mathbb{D}\}$ be the collection of all dissipative sequences for T. Then,*

(i) $\mathbb{C} \in \mathscr{D}, \ \mathbb{D} \in \mathscr{D} \implies \mathbb{C} \cup \mathbb{D} \in \mathscr{D}$.
(ii) $\mathbb{D} \in \mathscr{D}, \ \mathbb{C} \subset \mathbb{D} \implies \mathbb{C} \in \mathscr{D}$.
(iii) $\mathbb{D} \in \mathscr{D} \implies \mathbb{D} + \{k\} \in \mathscr{D}$ *for every* $k \in \mathbb{Z}$.
(iv) $\mathbb{D} = \{d_i : i = 1, 2, \ldots\} \in \mathscr{D} \implies \lim_{i \to \infty} \frac{i}{d_i} = 0$.

We first prove the following two lemmas.

Lemma 5.2.6. *Let T be an infinite ergodic transformation on $(X, \mathscr{B}, m)$, then*

$$m(A) < \infty \implies \lim_{n \to \infty} \frac{1}{n} m\left(\bigcup_{i=0}^{n} T^i A \right) = 0.$$

Proof. Let $A_0 = A$, and for each $k = 1, 2, \ldots$ let $A_k = (\bigcup_{i=0}^{k} T^i A) \smallsetminus (\bigcup_{i=0}^{k-1} T^i A)$. Then, for each $n \geq 0$, we have $\bigcup_{i=0}^{n} T^i A = \bigcup_{k=0}^{n} A_k \ (disj)$. We note that

$$m(A_k) = m\left[T^k A \smallsetminus \left(\bigcup_{i=0}^{k-1} T^i A \cap T^k A\right)\right] = m\left(A \smallsetminus \left(\bigcup_{j=1}^{k} T^{-j} A \cap A\right)\right)$$

$$= m(A) - m\left(\bigcup_{j=1}^{k} T^{-j} A \cap A\right).$$

T ergodic and $m(A) > 0 \implies \bigcup_{j=1}^{k} T^{-j} A \cap A \nearrow A$ as $k \to \infty$.
Therefore

$$\lim_{k \to \infty} m(A_k) = m(A) - m(A) = 0,$$

which implies

$$\lim_{n \to \infty} \frac{1}{n} m\left(\bigcup_{i=0}^{n} T^i A\right) = \lim_{n \to \infty} \frac{1}{n} \sum_{k=0}^{n} m(A_k) = 0. \qquad \square$$

Lemma 5.2.7. *If* $\{n_i : i \geq 1\}$ *is a sequence of positive integers having a positive upper density, then for any set* A *of finite measure,* $\liminf_{k \to \infty} \frac{1}{k} m \left(\bigcup_{i=1}^{k} T^{n_i} A\right) = 0$.

Proof. Let $\limsup_{k \to \infty} \dfrac{k}{n_k} = \alpha > 0$ for the sequence $\{n_i\}$. Then, $\liminf_{k \to \infty} \dfrac{n_k}{k} = \dfrac{1}{\alpha} < \infty$.
Therefore,

$$\liminf_{k \to \infty} \frac{1}{k} m\left(\bigcup_{j=1}^{k} T^{n_i} A\right) \leq \liminf_{k \to \infty} \left(\frac{n_k}{k} \cdot \frac{1}{n_k} m\left(\bigcup_{j=1}^{n_k} T^j A\right)\right)$$

$$\leq \liminf_{k \to \infty} \frac{n_k}{k} \cdot \limsup_{k \to \infty} \frac{1}{n_k} m\left(\bigcup_{j=1}^{n_k} T^j A\right)$$

$$= \frac{1}{\alpha} \cdot 0 = 0,$$

since the last lim sup equals 0 by Lemma 5.2.6. $\qquad \square$

Proof (of Proposition 5.2.5). (i), (ii), and (iii) follow from the definitions.

To show (iv), it suffices to show that if an infinite sequence of integers $\{n_i : i = 1, 2, \ldots\}$ has a positive upper density (i.e., $\limsup_{i \to \infty} \frac{i}{n_i} > 0$), then there exists a function $f \in L^1(m)$ with $f \geq 0$ a.e., for which $\sum_{i=1}^{\infty} f(T^{n_i}(x)) = \infty$ on a set A of positive measure.

Let us consider a sequence $\{n_i : i = 1, 2, \ldots\}$ with positive upper density, and let A be a set with $0 < m(A) < \infty$. In view of Lemma 5.2.6, we can get a sequence $k_1 < k_2 < \cdots < k_p < \cdots$ of integers such that, if $A(p) = \bigcup_{i=1}^{k_p} T^{n_i} A$, then

$$\frac{1}{k_p} m(A(p)) < \frac{1}{p} \quad \text{for } p = 1, 2, \ldots.$$

Next we define a function f as follows:

$$f(x) = \sum_{p=1}^{\infty} \frac{\chi_{A(p)}(x)}{p^2 m(A(p))}.$$

Clearly, $f(x) \geq 0$ a.e., and

$$\int_X f(x) \, dm(x) = \sum_{p=1}^{\infty} \frac{1}{p^2} < \infty.$$

Hence $f \in L^1(m)$. On the other hand, for $x \in A$ we have

$$\sum_{j=1}^{k_p} \chi_{A(p)}(T^{n_j} x) \geq k_p \quad \text{for each } p.$$

Therefore for every $x \in A$,

$$\sum_{j=1}^{\infty} f(T^{n_j} x) = \sum_{p=1}^{\infty} \sum_{j=1}^{\infty} \frac{\chi_{A(p)}(T^{n_j} x)}{p^2 m(A(p))} \geq \sum_{p=1}^{\infty} \sum_{j=1}^{k_p} \frac{\chi_{A(p)}(T^{n_j} x)}{p^2 m(A(p))}$$

$$\geq \sum_{p=1}^{\infty} \frac{k_p}{p^2 m(A(p))} \geq \sum_{p=1}^{\infty} \frac{1}{p} = \infty,$$

as $\dfrac{k_p}{m(A(p))} > p$ for each p. Since $m(A) > 0$, this shows that the sequence $\{n_i\}$ is not a dissipative sequence. $\square$

As we saw in Corollary 5.2.3, every ww sequence for an infinite ergodic transformation T is a dissipative sequence for T. It also follows, from the definition of dissipative sequences, that the union of two dissipative sequences is again a dissipative sequence for T. In the following remark, however, we show that the union of two ww sequences for an infinite ergodic transformation T in general is not a ww sequence for T.

Remark 5.2.8. We note that if an infinite sequence $\mathbb{W} = \{w_i\}$ is a ww sequence for an ergodic transformation, then $\mathbb{W} - \mathbb{W}$ cannot contain the set of all integers $\mathbb{Z}$. Next, let us recall the construction of the infinite ergodic transformation T in the First Basic Example in the previous chapter. There, the transformation T was constructed by extending the finite ergodic transformation S defined on the measure space $(X_0, \mathscr{B}_0, m_0)$, and using a sequence of $\{f(n)\}$ isomorphic copies of some subsets $\{A_n\}$ of X_0. By carefully increasing the sequence of integers $\{f(n)\}$ it is possible to construct another infinite ergodic transformation T that admits the following two ww sequences: an infinite sequence $\mathbb{W} = \{w_i\}$ and another infinite sequence $\mathbb{V} = \{v_i\}$, where $v_i = w_i + i$ for $i = 1, 2, \ldots$. It follows then for this transformation T that the sequence $\mathbb{U} = \mathbb{W} \cup \mathbb{V}$ has the property that $\mathbb{U} - \mathbb{U} \supset \mathbb{Z}$. This shows that the sequence $\mathbb{W} \cup \mathbb{V}$ cannot be a ww sequence for the transformation T; see [30].

In [47] it is shown that every dissipative sequence for an infinite ergodic transformation has density 0; moreover, an interesting example of an infinite ergodic transformation T is constructed that possesses a dissipative sequence that is not the union of a finite number of its ww sequences.

5.3 The Sequences in a Different Setting

In this section we describe the collections $\mathscr{W}$, $\mathscr{D}$, $\mathscr{R}$, etc. of all weakly wandering, dissipative, recurrent and related sequences, respectively for an infinite ergodic transformation T in a different setting following [27].

We consider the space of ultrafilters of subsets of the integers $\mathbb{Z}$.

By an *ultrafilter* $\mathfrak{p}$ of subsets $\mathbb{E}$ of the integers $\mathbb{Z}$ we mean a non-empty collection of non-empty subsets $\mathbb{E}$ of $\mathbb{Z}$ with the following two properties:

(i) $\mathbb{E}_1 \in \mathfrak{p}$, $\mathbb{E}_2 \in \mathfrak{p} \implies \mathbb{E}_1 \cap \mathbb{E}_2 \in \mathfrak{p}$.
(ii) For every non-empty subset $\mathbb{E} \subset \mathbb{Z}$, either $\mathbb{E} \in \mathfrak{p}$ or $\mathbb{E}^c \in \mathfrak{p}$.

It is not difficult to show that for an ultrafilter $\mathfrak{p}$ only one of the following holds:

(a) There exists a number $p \in \mathbb{Z}$ such that $\bigcap_{\mathbb{E} \in \mathfrak{p}} \mathbb{E} = \{p\}$.

(b) $\bigcap_{\mathbb{E} \in \mathfrak{p}} \mathbb{E} = \emptyset$.

We say that an ultrafilter $\mathfrak{p}$ is *principal* if (a) holds, and it is *free* if (b) holds. It can be shown that an ultrafilter $\mathfrak{p}$ is a principal ultrafilter corresponding to the number $p \in \mathbb{Z}$ if and only if $\mathfrak{p} = \{\mathbb{E} \subset \mathbb{Z} : p \in \mathbb{E}\}$ and $\mathfrak{p}$ is a free ultrafilter if and only if it contains no finite subsets of $\mathbb{Z}$.

It is customary to denote the set of all ultrafilters of $\mathbb{Z}$ by $\beta\mathbb{Z}$. For a non-empty subset $\mathbb{E}$ of $\mathbb{Z}$ let us define the set

$$\mathscr{O}(\mathbb{E}) = \{\mathfrak{p} \in \beta\mathbb{Z} : \mathbb{E} \in \mathfrak{p}\}.$$

It is clear that if $\mathbb{E} \subset \mathbb{F}$, then $\mathcal{O}(\mathbb{E}) \subset \mathcal{O}(\mathbb{F})$. Suppose for a pair $\mathbb{E}_1, \mathbb{E}_2$ of non-empty subsets of $\mathbb{Z}$, $\mathfrak{p} \in \mathcal{O}(\mathbb{E}_1) \cap \mathcal{O}(\mathbb{E}_2)$, then $\mathbb{E}_1 \in \mathfrak{p}$ and $\mathbb{E}_2 \in \mathfrak{p}$. By property (i) above we have $\mathfrak{p} \in \mathcal{O}(\mathbb{E}_1 \cap \mathbb{E}_2) \subset \mathcal{O}(\mathbb{E}_1) \cap \mathcal{O}(\mathbb{E}_2)$. This implies that the family $\{\mathcal{O}(\mathbb{E}) : \emptyset \neq \mathbb{E} \subset \mathbb{Z}\}$ is a basis for a topology on $\beta\mathbb{Z}$. We regard the space $\beta\mathbb{Z}$ as a topological space with the topology given by this basis.

By identifying each point $p \in \mathbb{Z}$ with the principal ultrafilter $\mathfrak{p}$ associated with p, that is, an ultrafilter $\mathfrak{p}$ for which $\bigcap_{E \in \mathfrak{p}} E = \{p\}$, we can identify the integers $\mathbb{Z}$ with the principal ultrafilters of $\beta\mathbb{Z}$. It is well-known that the space $\beta\mathbb{Z}$ is compact with respect to this topology, and $\mathbb{Z}$ regarded as a subset of $\beta\mathbb{Z}$ is a dense subset of it. In fact, $\beta\mathbb{Z}$ with this topology is the Stone–Čech compactification of the discrete topological space $\mathbb{Z}$.

In the sequel we consider the subset of $\beta\mathbb{Z}$ consisting of all free ultrafilters, and denote it by $\mathcal{F}$. We regard $\mathcal{F}$ as a topological space with the relative topology inherited from the Stone–Čech compactification $\beta\mathbb{Z}$. Then, in the space $\mathcal{F}$ we have $\mathcal{O}(E) \subset \mathcal{O}(F)$ if and only if $|\mathbb{E} \cap \mathbb{F}^c| < \infty$, and $\mathcal{O}(E) \cap \mathcal{O}(F) = \emptyset$ if and only if $|\mathbb{E} \cap \mathbb{F}| < \infty$.

For any subset $\mathcal{C}$ of the space $\mathcal{F}$, let us denote by $\mathcal{C}^*$ the set $(\overline{\mathcal{C}})^\circ$, namely interior of the closure of the set $\mathcal{C}$. $\mathcal{C}^*$ is called the *regularization* of $\mathcal{C}$. An open subset $\mathcal{O}$ of $\mathcal{F}$ is called a *regularly open set* if $\mathcal{O}^* = \mathcal{O}$. Let us also denote by $\mathcal{C}^\sharp$ the set $(\overline{\mathcal{C}})^c$, namely complement of the closure of the set $\mathcal{C}$. $\mathcal{C}^\sharp$ is called the *exterior* of $\mathcal{C}$.

Let T be an ergodic measure-preserving transformation defined on the infinite measure space $(X, \mathcal{B}, m)$. Let us consider the following collections of subsets of $\mathbb{Z}$:

$$\mathcal{W} = \{\mathbb{W} \subset \mathbb{Z} : \mathbb{W} \text{ is a } ww \text{ sequence for } T\}.$$
$$\mathcal{R} = \{\mathbb{R} \subset \mathbb{Z} : |\mathbb{R} \cap \mathbb{W}| < \infty \text{ for all } \mathbb{W} \in \mathcal{W}\}.$$
$$\mathcal{V} = \{\mathbb{V} \subset \mathbb{Z} : |\mathbb{V} \cap \mathbb{R}| < \infty \text{ for all } \mathbb{R} \in \mathcal{R}\}.$$

Remark 5.3.1. We notice the following:

- By Definition 3.3.1, $\mathcal{R}$ is the collection of all recurrent sequences for T.
- From Lemma 3.2.5 and Proposition 3.3.2 follows that $\mathcal{V}$ is the set of all sequences $\mathbb{V} = \{v_i\}$ with the property that $\liminf_{i \to \infty} m(T^{v_i} A \cap A) = 0$ for all A with $m(A) < \infty$.
- Let us consider the collection $\mathcal{R}' = \{\mathbb{R}' \subset \mathbb{Z} : |\mathbb{R}' \cap \mathbb{V}| < \infty \text{ for all } \mathbb{V} \in \mathcal{V}\}$. We note that since $\mathcal{W} \subset \mathcal{V}$ we have for $\mathbb{R}' \in \mathcal{R}'$, $|\mathbb{R}' \cap \mathbb{W}| < \infty$ for all $\mathbb{W} \in \mathcal{W}$, and this implies $\mathcal{R}' = \mathcal{R}$.
- Let us also consider the following collection of subsets of $\mathbb{Z}$: $\mathcal{U} = \{\mathbb{U} = \{u_i\} \subset \mathbb{Z} : \limsup_{i \to \infty} m(T^{u_i} A \cap A) > 0 \text{ for some } A \text{ with } m(A) < \infty\}$. By Corollary 3.3.4 we have $\mathcal{U} \supset \mathcal{R}$.
- Consider $\mathcal{D}$ the collection of dissipative sequences for T (see Definition 5.2.1). Then by Corollary 5.2.3 and Proposition 5.2.4, $\mathcal{W} \subset \mathcal{D} \subset \mathcal{V}$.

For a subset $\mathbb{E} \subset \mathbb{Z}$ let us recall the open set $\mathcal{O}(\mathbb{E}) = \{\mathfrak{p} \in \mathcal{F} : \mathbb{E} \in \mathfrak{p}\}$. For any collection $\mathcal{E}$ of subsets $\mathbb{E} \subset \mathbb{Z}$ let us write $\mathcal{O}(\mathcal{E}) = \bigcup_{\mathbb{E} \in \mathcal{E}} \mathcal{O}(\mathbb{E})$. Then, from the last two items of Remark 5.3.1 above we conclude that $\mathscr{O}(\mathscr{R}) \subset \mathscr{O}(\mathscr{U})$ and $\mathscr{O}(\mathscr{W}) \subset \mathscr{O}(\mathscr{D}) \subset \mathscr{O}(\mathscr{V})$.

In the following theorem we describe various properties of several open sets of $\mathfrak{F}$.

Theorem 5.3.2. *Let T be an ergodic measure-preserving transformation defined on the infinite measure space $(X, \mathscr{B}, m)$, and let $\mathscr{W} = \{W\}, \mathscr{R} = \{\mathbb{R}\}, \mathscr{V} = \{V\}$ be the collections of subsets of $\mathbb{Z}$ as described above. Then*

(i) *$\mathscr{O}(\mathscr{W})^{\#} = \mathscr{O}(\mathscr{R}) = \mathscr{O}(\mathscr{R})^*$. $\mathscr{O}(\mathscr{R})$ is the exterior of $\mathscr{O}(\mathscr{W})$, and $\mathscr{O}(\mathscr{R})$ is a regularly open set.*

(ii) *$\mathscr{O}(\mathscr{R})^{\#} = \mathscr{O}(\mathscr{V}) = \mathscr{O}(\mathscr{V})^*$. $\mathscr{O}(\mathscr{V})$ is the exterior of $\mathscr{O}(\mathscr{R})$, and $\mathscr{O}(\mathscr{V})$ is a regularly open set.*

(iii) *$\mathscr{O}(\mathscr{W})^* = \mathscr{O}(\mathscr{V})$. $\mathscr{O}(\mathscr{V})$ is the regularization of $\mathscr{O}(\mathscr{W})$.*

For a collection $\mathscr{K} = \{\mathbb{K}\}$ of subsets of $\mathbb{Z}$ let us write $\mathscr{K}^{\perp} = \{\mathbb{E} \subset \mathbb{Z} : |\mathbb{E} \cap \mathbb{K}| < \infty$ for all $\mathbb{K} \in \mathscr{K}\}$. First we prove the following lemma.

Lemma 5.3.3. $\mathscr{O}(\mathscr{K}^{\perp}) = \mathscr{O}(\mathscr{K}^{\perp})^* = \mathscr{O}(\mathscr{K})^{\#}$.

Proof. From the definition of $\mathscr{K}^{\perp}$ follows: $\mathscr{O}(\mathscr{K}) \cap \mathscr{O}(\mathscr{K}^{\perp}) = \emptyset$, which implies $\mathscr{O}(\mathscr{K}) \cap \mathscr{O}(\mathscr{K}^{\perp})^* = \emptyset$. Then we conclude

$$\mathscr{O}(\mathscr{K}^{\perp}) \subset \mathscr{O}(\mathscr{K}^{\perp})^* \subset \mathscr{O}(\mathscr{K})^{\#}.$$

For an ultrafilter $\mathfrak{p} \in \mathscr{O}(\mathscr{K})^{\#}$ $\exists$ $\mathbb{E} \subset \mathbb{Z}$ such that $\mathfrak{p} \in \mathscr{O}(\mathbb{E}) \subset \mathscr{O}(\mathscr{K})^{\#}$, which implies $|\mathbb{E} \cap \mathbb{K}| < \infty$ for all $\mathbb{K} \in \mathscr{K}$.

Therefore $\mathbb{E} \in \mathscr{K}^{\perp}$ and $\mathfrak{p} \in \mathscr{O}(\mathbb{E}) \subset \mathscr{O}(\mathscr{K}^{\perp})$. Thus $\mathscr{O}(\mathscr{K})^{\#} \subset \mathscr{O}(\mathscr{K}^{\perp})$, and this completes the proof. $\square$

Proof (Theorem 5.3.2). We let $\mathscr{K} = \mathscr{W}$ in Lemma 5.3.3 and note that $\mathscr{K}^{\perp} = \mathscr{R}$. Then Lemma 5.3.3 implies (i).

We let $\mathscr{K} = \mathscr{R}$ in Lemma 5.3.3 and note that $\mathscr{K}^{\perp} = \mathscr{V}$. Then Lemma 5.3.3 implies (ii).

From (i) and (ii) we get: $\mathscr{O}(\mathscr{W})^{\#\#} = \mathscr{O}(\mathscr{R})^{\#} = \mathscr{O}(\mathscr{V})$.

Since for any set of ultrafilters $\mathscr{E} \subset \mathscr{F}$, we have $\mathscr{E}^* \supset \mathscr{E}^{\#\#}$, we obtain from the above $\mathscr{O}(\mathscr{W})^* \supset \mathscr{O}(\mathscr{V})$. Also $\mathscr{W} \subset \mathscr{V}$ implies $\mathscr{O}(\mathscr{W})^* \subset \mathscr{O}(\mathscr{V})^* = \mathscr{O}(\mathscr{V})$. Therefore $\mathscr{O}(\mathscr{W})^* = \mathscr{O}(\mathscr{V})$. $\square$

Chapter 6
Isomorphism Invariants

In this chapter we examine several isomorphism invariants associated to an infinite ergodic transformation. We use these invariants to show the non-isomorphism of various infinite ergodic transformations. In Sect. 6.1 the role of *eww* sets and sequences is discussed and used to distinguish between some infinite ergodic transformations. In Sect. 6.2 it is shown that there exists a class of infinite ergodic transformations that exhibit a regularity in the size of the return sets of finite measure. We define (an isomorphism invariant) the α-type for such transformations. Previously, in Sect. 3.3 recurrent sequences for an infinite ergodic transformation were defined, and then in Sect. 4.1 the First Basic Example was constructed and all its recurrent sequences were computed. This is extended in Sect. 6.3 to a family of transformations where all the recurrent sequences for the members of that family are computed and shown to distinguish between any two members of the family. Finally, in Sect. 6.4 the growth distribution of *ww* sets is also shown to be an effective isomorphism invariant.

In each of the sections of this chapter several infinite ergodic transformations are constructed to exhibit the effectiveness of the various isomorphism invariants that are being discussed. It is interesting to note that all of these examples are variations of the First Basic Example.

6.1 Exhaustive Weakly Wandering Sets

Definition 6.1.1. Two measurable and nonsingular transformations, T defined on $(X, \mathcal{B}, \mu)$ and S defined on $(Y, \mathcal{C}, \nu)$, are *isomorphic* if there exists an invertible map $\phi : Y \longrightarrow X$ such that $\phi \circ S = T \circ \phi$ and $\nu(\phi^{-1}(B)) = \mu(B)$ for all $B \in \mathcal{B}$.

In this section we present two infinite ergodic transformations that have an *eww* sequence in common, yet the transformations are not isomorphic. The subtlety arises in the analysis of the *eww* sets and how they "sit."

© Springer Japan 2014
S. Eigen et al., *Weakly Wandering Sequences in Ergodic Theory*,
Springer Monographs in Mathematics, DOI 10.1007/978-4-431-55108-9_6

As defined in Chap. 5 an infinite ergodic transformation T is of *finite type* if it admits an *eww* set of finite measure.

Proposition 6.1.2. *Let T defined on $(X, \mathscr{B}, \mu)$ be an infinite ergodic transformation of finite type and S defined on $(Y, \mathscr{C}, \nu)$ be another ergodic measure-preserving transformation. Then*

(P1) Every eww sequence for T is also an eww sequence for the transformation $T \times S$ defined on $(X \times Y, \mathscr{B} \times \mathscr{C}, \mu \times \nu)$.

(P2) If S preserves an infinite measure ν then the transformations T and $T \times S$ are not isomorphic.

(P3) If the transformations T and S are isomorphic, then every eww sequence for T is also an eww sequence for S.

Proof. Let W be an *eww* set for T with the *eww* sequence $\{n_i : i \geq 1\}$.

Then $X = \bigcup_{i=1}^{\infty} T^{n_i} W$ *(disj)* implies $X \times Y = \bigcup_{i=1}^{\infty} (T \times S)^{n_i} (W \times Y)$ *(disj)*, and this proves *(P1)*.

Next we let the *eww* set W have finite measure. Suppose there exists an isomorphism $\phi : X \times Y \longrightarrow X$ between the transformations T and $T \times S$. Then the sets $\phi^{-1} W$ and $W \times Y$ are both *eww* sets for the transformation $T \times S$ with the same *eww* sequence $\{n_i : i \geq 1\}$ mentioned above. However, $(\mu \times \nu)(\phi^{-1} W) < \infty$, while $(\mu \times \nu)(W \times Y) = \infty$. This is a contradiction to Theorem 5.1.2 and proves *(P2)*.

The proof of *(P3)* is similar to the proof of *(P1)*. Namely,

$$X = \bigcup_{i=1}^{\infty} T^{n_i} W \text{ (disj) implies } Y = \bigcup_{i=1}^{\infty} S^{n_i} \phi^{-1} W \text{ (disj)}. \qquad \square$$

In analyzing the *eww* sets for the product transformation there are two cases of interest. The transformation S defined on Y may be finite measure-preserving on a probability space, or S may be infinite measure-preserving. When S is finite measure-preserving then the measure of the *eww* set W is the same as the product measure of the *eww* set $W \times Y$. Hence it could be possible for T and $T \times S$ to be isomorphic. However, property *(P2)* of Proposition 6.1.2 addresses the above when T is an infinite ergodic transformation of finite type as defined in Sect. 5.1.

Technically, in Proposition 6.1.2 property *(P1)* does not follow from property *(P3)*. This is because the projection map from $X \times Y$ to X is not an isomorphism when the measure of Y is infinite. This is related to the issue of factors of infinite measure-preserving transformations. In the above situation one would expect to be able to say that T is a factor of $T \times S$. Other authors (Aaronson [1]) deal with this in a different way. A common method to overcome the problem with Y being of infinite measure is to switch to an equivalent finite measure as in Chap. 1, then find an invariant sub-σ-algebra of sets and restrict T to this σ-algebra. Finally, convert the finite measure back to an invariant infinite measure. However, this is not always possible. In particular, let us consider when T is the Second Basic Example of Chap. 4. Recall that Q is the measure-doubling transformation which commutes with T. The invariant sets for Q, $\{A \in \mathscr{B} : Q(A) = A\}$, give an invariant

σ-algebra $\mathscr{I} \subset \mathscr{B}$ for T. However, no σ-finite measure invariant for T can be put on this because T has no equivalent invariant measure as it comes from a type III transformation.

In the case when S is finite measure-preserving, it is still possible for T and $T \times S$ to be non-isomorphic even though they have many *eww* sequences in common. Here are two examples (the proofs use tools which we do not pursue in this monograph).

Let us consider the First Basic Example T defined on $(X, \mathscr{B}, m)$ as given in Chap. 4. Let S defined on $(Y, \mathscr{C}, \nu)$ be a positive entropy transformation which is finite measure-preserving and strongly mixing. Consider $T \times S$ on the product measure space $(X \times Y, \mathscr{B} \times \mathscr{C}, m \times \nu)$. Since S is strongly mixing, the product $T \times S$ is an infinite ergodic transformation. As before, every *eww* sequence $\{n_i\}$ for T is also an *eww* sequence for $T \times S$. Entropy considerations show that T and $T \times S$ are not isomorphic. Specifically, the transformation T induces on every set of finite measure a transformation with zero entropy. However $T \times S$ induces on $X_0 \times Y$ a transformation with positive entropy (where X_0 is the set of non-dyadic rationals in $[0, 1]$). Hence the pair of transformations T and $T \times S$ cannot be isomorphic.

Another example also begins with the First Basic Example T of Chap. 4. We let S defined on $\{0, 1, 2\}$ be the cyclic permutation $0 \mapsto 1 \mapsto 2 \mapsto 0$, with the atomic measure $(1/3, 1/3, 1/3)$. It is clear that $T \times S$ is ergodic by analyzing the map on the product of cylinder sets crossed with $\{i\}$ for $i = 0, 1, 2$. Examining the two maps we see that $e^{2\pi/3}$ is in the L^∞-spectrum of $T \times S$ but not in the L^∞-spectrum of T; again the two maps are not isomorphic.

In an isomorphism of two infinite ergodic transformations an *eww* set maps to an *eww* set. Using induced transformations, we state an elementary isomorphism theorem. First we recall some notation. Given a transformation T defined on $(X, \mathscr{B}, \mu)$ and a set $A \in \mathscr{B}$ of positive measure, denote by T_A the induced transformation on the set A as defined in Chap. 4. For each point $x \in A$ we also denote by $r_A(x) = \{n > 0 : T^n x \in A, T^i x \notin A, 0 < i < n\}$ the return times to A.

Proposition 6.1.3. *Let two infinite ergodic transformations T defined on $(X, \mathscr{B}, \mu)$, S defined on $(Y, \mathscr{C}, \nu)$ and sets $B \in \mathscr{B}$, $C \in \mathscr{C}$ with $\mu(B) = \nu(C) > 0$ be given. Suppose the two induced transformations T_B and S_C are isomorphic via ϕ, that is $\phi \circ S_C(y) = T_B \circ \phi(y)$ for all $y \in C$; and assume $r_C(y) = r_B(\phi(y))$. Then S and T are isomorphic.*

Proof. The isomorphism ϕ can be extended to the whole space. Suppose $r_C(y) = r_B(\phi(y)) = n$. Then there are $n - 1$ points $\{Sy, S^2y, S^3y, \ldots, S^{n-1}y\}$ not in C, and $n - 1$ points $\{T(\phi y), T^2(\phi y), \ldots, T^{n-1}(\phi y)\}$ not in B. We extend ϕ to map these points to each other in the given order. $\square$

From the above we get a necessary condition: partition the sets
$$B = \bigcup_{i=1}^{\infty} B_i \quad \text{where} \quad B_i = \{x \in B : T^i x \in B, T^j x \notin B, 0 < j < i\}$$
and
$$C = \bigcup_{i=1}^{\infty} C_i \quad \text{where} \quad C_i = \{x \in C : S^i x \in C, S^j x \notin C, 0 < j < i\}.$$

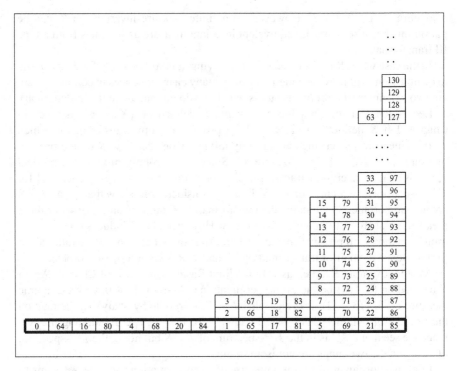

Fig. 6.1   Initial *eww* set X_0 for the First Basic Example

Corollary 6.1.4. *A necessary condition for the existence of an isomorphism of the two transformations T and S which maps the set B to C is $\mu(B_i) = \nu(C_i)$ for $i \geq 1$.*

Next we illustrate the above for the First Basic Example.

The measure of the sets A_n for the First Basic Example (those points in A which first come back in n steps) can be derived from the construction in Fig. 6.1. In particular, $m(A_n) = 0$ except for the following values of $n = 1+2+2^3+\cdots+2^{2k-1}$.

$$m(A_1) = 1/2$$

$$m(A_3) = 1/4$$

$$m(A_{11}) = 1/8$$

$$\vdots$$

$$m(A_{1+2+2^3+\cdots+2^{2k-1}}) = 1/2^{k+1}$$

$$\vdots$$

The point of Corollary 6.1.4 is that it is possible for different *eww* sets to "sit differently" in a transformation even when they have the same *eww* sequences and isomorphic induced transformations. We illustrate this as follows.

We let T defined on $(X, \mathcal{B}, m)$ be the First Basic Example of Chap. 4. We recall that $X_0 = \{x \in (0, 1), x \neq \text{dyadic rational}\}$ is an *eww* set with the *eww* sequence $\{n_i\}$, where

$$n_i = \varepsilon_0 2^1 + \varepsilon_1 2^3 + \cdots + \varepsilon_k 2^{2k+1} \quad \text{if} \quad i = \varepsilon_0 2^0 + \varepsilon_1 2^1 + \cdots + \varepsilon_k 2^k,$$

$\varepsilon_j = 0$ or 1 for $j = 0, 1, \ldots, k$ and $k \geq 1$.

For the remainder of this chapter let us use the notation $\dot{\cup}$ to denote the disjoint union of sets. Namely, $A \dot{\cup} B$ will denote the set $A \cup B \, (disj)$.

Next, let us partition the set $X_0 = A \dot{\cup} B$, where $A = \{x \in X_0, \ 0 < x < 1/2\}$ and $B = \{x \in X_0, \ 1/2 < x < 1\}$. By the definition of the transformation T we have $TA = B$. Hence,

$$X = \dot{\bigcup}_{i=1}^{\infty} T^{n_i}(X_0) = \dot{\bigcup}_{i=1}^{\infty} T^{n_i}(A \dot{\cup} TA) = \dot{\bigcup}_{i=1}^{\infty} T^{n_i} A \ \dot{\cup} \ \dot{\bigcup}_{i=1}^{\infty} T^{n_i} TA. \quad (6.1)$$

Proposition 6.1.5. *The set $X_1 = A \dot{\cup} T^2 A$ is another eww set for the transformation T of the First Basic Example with the same sequence $\{n_i\}$ as given above.*

Proof. Applying T to (6.1) gives

$$TX = X = \dot{\bigcup}_{i=1}^{\infty} T^{n_i} TA \ \dot{\cup} \ \dot{\bigcup}_{i=1}^{\infty} T^{n_i} T^2 A.$$

Therefore

$$\dot{\bigcup}_{i=1}^{\infty} T^{n_i} A = \dot{\bigcup}_{i=1}^{\infty} T^{n_i} T^2 A.$$

$\square$

Figures 6.1 and 6.2 respectively show the original *eww* set and the new *eww* set as they "sit" in the skyscraper representation of the transformation.

Proposition 6.1.6. *The transformations T_{X_0} and T_{X_1} induced by the transformation T of the First Basic Example on the sets X_0 and X_1 are isomorphic.*

Proof. We note that for $x \in A$, $T_{X_1}(x) = T^2(T_{X_0}(x))$, and for $x \in B$, $T_{X_0}(x) = T_{X_1}(x)$. Then the map $\phi(x) = \begin{cases} x & \text{if } x \in A, \\ T^2 x & \text{if } x \in B \end{cases}$ gives the isomorphism. $\square$

This isomorphism does not satisfy Proposition 6.1.3 because the corresponding return time functions r_{X_0} and r_{X_1} do not match. For example, each $x \in A$ has $r_{X_0}(x) = 1$ and $r_{X_1}(x) = 3$. However, there still might be another isomorphism to which the proposition could apply. Using Corollary 6.1.4 we can see that the isomorphism of the induced maps on X_0 and X_1 does not extend to an isomorphism of the entire space.

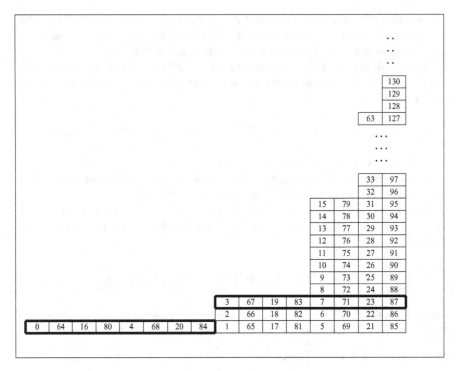

Fig. 6.2 A second *eww* set X_1 for the First Basic Example

Proposition 6.1.7. *There is no self-isomorphism of the transformation T of the First Basic Example defined on $(X, \mathscr{B}, m)$ which maps the set X_0 to X_1.*

Proof. The set $\{x \in X_0 : Tx \in X_0\}$ has measure $1/2$. The set $\{x \in X_1 : Tx \in X_1\}$ has measure $1/4$. The result follows by Corollary 6.1.4.   □

6.2 α-Type Transformations

We recall that the transformation T in the First Basic Example of Chap. 4 was constructed as a skyscraper transformation over the von Neumann transformation, as illustrated in Fig. 4.2 as well as in Figs. 6.1 and 6.2.

Yet another construction of the transformation T is as a rank-one cutting and stacking map. This is revealed in Eqs. (4.5) and (4.6). In particular, the sets C_i as defined in the skyscraper construction are columns of height 4^i with the map T moving up the columns linearly.

The following is the cutting and stacking construction for the First Basic Example. We begin with the set $X_0 = [0, 1)$, the unit interval with the dyadic rationals removed. We define the column C_0 of height $h_0 = 1$ to be the set X_0, and we consider the interval $[1, \infty)$ (minus the dyadic rationals) as our spacer set. We cut the column C_0 into two equal halves, then place the right subcolumn over the left subcolumn and add two spacer intervals (measure 1/2) on the top. This produces a new column C_1 of height $h_1 = 4$. We repeat this process by induction. Column C_n of height $h_n = 4^n$ is cut in half. The right subcolumn is placed above the left subcolumn. Then $2 \cdot h_n$ spacer intervals are placed on the top of this column to create the next column C_{n+1}.

The *eww* sequence $\{n_i\}$ for the *eww* set X_0 is given by

$$n_i = \varepsilon_0 2^1 + \varepsilon_1 2^3 + \cdots + \varepsilon_k 2^{2k+1} \quad \text{if} \quad i = \varepsilon_0 2^0 + \varepsilon_1 2^1 + \cdots + \varepsilon_k 2^k,$$

$\varepsilon_j = 0$ or 1 for $j = 0, 1, \ldots, k$ and $k \geq 1$.

We generalize the First Basic Example as follows: Let $p \geq 3$ be an integer. We obtain T_p as a skyscraper over a p-adic odometer via the following cutting and stacking rank-one construction.

Let X_p be the unit interval $[0, 1)$ with the p-adic rationals removed (as before, this is a denumerable set). The set X_p is the column C_0 of height $h_0 = 1$. At the n-th stage we will have a column C_n of height $h_n = p^{2n}$. We divide the column into p equal width subcolumns. Stack the columns, right atop left then add $(p - 1)ph_n$ spacer intervals. This gives a column C_{n+1} of height $h_{n+1} = p^{2n+2}$. This process defines T_p as an infinite ergodic transformation whose induced transformation on X_p is the p-adic odometer.

As in the case of the First Basic Example we define a sequence of integers which will be an *eww* sequence for T_p with X_p as the *eww* set.

For $i = \sum_j^k \varepsilon_j p^j$ a finite sum, $\varepsilon_j \in \{0, 1, \ldots, p-1\}$, we define the sequence of integers $\{n_i\}$ by

$$\mathbb{P}(p) = \left\{ n_i : n_i = \sum_{j=0}^k \varepsilon_j p^{2j+1}, \text{ where } i = \sum_{j=0}^k \varepsilon_j p^j, \varepsilon_j \in \{0, 1, \ldots, p-1\} \right\}.$$

$$(6.2)$$

For each of these transformations T_p we have an *eww* sequence $\mathbb{P}(p)$. These *eww* sequences are clearly different suggesting the transformations are non-isomorphic. However, transformations in general, and these in particular have many *eww* sequences and so it is not clear that they are not isomorphic. In this section we show the non-isomorphism of the family $\{T_p\}$ by a more detailed analysis of the size of the intersections $T^r X_p \cap X_p$.

We recall Theorem 4.1.1. This gives a complete description of the recurrent sequences for the First Basic Example. The description depends upon Eqs. (4.8). From this we see that when $p = 2$ the transformation T_2 is just the First Basic Example T, and for any set satisfying $0 < m(A) < \infty$ we have $\limsup_{n \to \infty} m(T^n A \cap A) = \frac{1}{2} m(A)$.

This motivates the following definition:

Definition 6.2.1. A transformation T is said to be of α-*type* for some $0 \leq \alpha \leq 1$ if for all sets A with $0 < m(A) < \infty$ we have

$$\limsup_{n \to \infty} m(T^n A \cap A) = \alpha m(A).$$

Clearly the α-type when it exists is an isomorphism invariant. For finite measure-preserving transformations the only α-type that can occur is $\alpha = 1$. In the infinite measure-preserving case, Hamachi and Osikawa showed there exist transformations of α-type for every $\alpha \in [0, 1]$; see [35]. Note further that not all infinite measure-preserving, ergodic transformations have to be of α-type for a single, fixed $\alpha > 0$ (see [35]).

The First Basic Example is of α-type for $\alpha = \frac{1}{2}$. The Third Basic Example was shown not to possess any recurrent sequences and so by definition is of α-type for $\alpha = 0$. In contrast the Second Basic Example possesses recurrent sequences (for example $\{2^k : k = 1, 2, \ldots\}$) but is not of α-type for any fixed α.

The following theorem shows that the maps T_p are all pairwise non-isomorphic.

Theorem 6.2.2. *The map T_p as defined previously is of α-type with $\alpha = (p-1)/p$.*

Proof. The proof is straightforward. We briefly outline the steps.

From the cutting, stacking and spacer construction of the map T_p it is easy to see that T_p satisfies the following, which is similar to Eq. (4.8).

$$m(T_p^n X_p \cap X_p) = (p-1)/p \text{ for } n = p^{2k},$$

$$m(T_p^n X_p \cap X_p) < (p-1)/p, \text{ for } n \neq p^{2k}. \tag{6.3}$$

This is then extended to any set which is a finite union of levels in a single column. By a standard approximation argument this extends to all sets of finite measure. □

As in the case of the First Basic Example it is possible to determine the size of $m(T_p^n X_p \cap X_p)$ for every $n = 1, 2, \ldots$; it is also possible to determine all recurrent sequences for T_p.

The following is easily seen in the cutting and stacking representation of T_p.
For $T = T_p$

$$m(T^n X_p \cap X_p) = \begin{cases} (p-1)/p & \text{if } n = p^{2k}, \\ (p-2)/p & \text{if } n = 2 \cdot p^{2k}, \\ \quad \vdots \\ 1/p & \text{if } n = (p-1)p^{2k}. \end{cases} \tag{6.4}$$

Next we consider the following sequence of integers: let $N_0 = \{0\}$, and for $k \geq 1$

$$N_k = \{n : n = \pm \varepsilon_1 p^{2q_1} \pm \varepsilon_2 p^{2q_2} \pm \cdots \pm \varepsilon_k p^{2q_k}, \ \varepsilon_i \in \{1, 2, \ldots, p-1\} \}, \quad (6.5)$$

where q_i are integers satisfying $0 \leq q_1 < q_2 < \cdots < q_k$.

Theorem 6.2.3. *The infinite set of integers* $\{r_i : i = 0, 1, 2, \ldots\}$ *is a recurrent sequence for the infinite ergodic transformation* T_p *if and only if there exist two positive integers* l *and* s *such that*

$$\{r_i : i = 0, 1, 2, \ldots\} \subset \bigcup_{n=-s}^{s} \bigcup_{k=0}^{l} (N_k + n). \quad (6.6)$$

As the proof of Theorem 6.2.3 is a direct extension of the proof of Theorem 4.1.1, we omit it.

6.3 Recurrent Sequences as an Isomorphism Invariant

In the previous section a family of transformations was given where each member of the family had a different α-type and thus any two of them were shown to be non-isomorphic. In this section a second family of transformations is presented where all the members of the family are of the same α-type with $\alpha = 1/2$. It is shown that these transformations are also, in general, not isomorphic to each other. The non-isomorphism is proved by showing that distinct transformations in the family have different recurrent sequences.

The details are similar to those presented in Sect. 4.1.2 where the First Basic Example was constructed and a complete description of all its recurrent sequences given. In this section the construction of Sect. 4.1.2 is slightly modified and a family of transformations is obtained where, similar to the First Basic Example, all the members in the family are of the same α-type with $\alpha = \frac{1}{2}$. The recurrent sequences for each member of the family are also described and used to show the non-isomorphism of the transformations. The proofs and constructions are essentially identical to those of Sect. 4.1.2.

We note that it is also possible to use techniques from finite ergodic theory (such as coding theory [51]) to prove the transformations are non-isomorphic, but we consider the recurrent sequences a more natural invariant for this infinite measure-preserving case.

6.3.1 Construction of the Transformation $T_{\bar{\varepsilon}}$

We recall from the previous section the cutting and stacking construction of the First Basic Example. We start with X_0, the unit interval with the dyadic rationals removed. At the nth stage the current column C_n is cut in half. On top of the right subcolumn 2^{2n+1} spacers are placed, and then the right subcolumn with its additional spacers is placed over the left subcolumn. The transformation is, as usual, extended by "going up" the new column. The variation in this section is that at each stage we make a choice of placing the 2^{2n+1} spacers either over the left subcolumn or over the right subcolumn before putting the right subcolumn above the left.

It is clear that, independent of the choice of placement of the spacers, the induced transformation defined on the set X_0 in each case is the same and is the von Neumann transformation S as presented in Sect. 4.1.2.

We first introduce some notation. Let $\bar{\varepsilon}$ be an infinite sequence of 0's and 1's, i.e. $\bar{\varepsilon} \in \prod_0^\infty \{0, 1\}$. For each $\bar{\varepsilon}$ we construct a transformation $T_{\bar{\varepsilon}}$. As was done earlier, each of these transformations is a simple variation of the First Basic Example; and in particular the First Basic Example is represented by $T_{\bar{\varepsilon}}$ where $\bar{\varepsilon} = (0, 0, \ldots)$.

As usual we start with the set X_0, the interval $[0, 1)$ with the dyadic rationals removed, and we consider X_0 as the column C_0 of height $h_0 = 1$ and width $w_0 = 1$. To construct the column C_1, a block of 2^1 spacers is added prior to stacking the right subcolumn over the left subcolumn. This block of spacers may be added either over the left or over the right subcolumn depending on the value of ε_0. If $\varepsilon_0 = 0$, the block of spacers is placed over the right subcolumn. If $\varepsilon_0 = 1$, the block of spacers is placed over the left subcolumn (see top of Fig. 6.3). In either case the column C_1 is of height $h_1 = 2^2$ and width $w_1 = 1/2$ with total measure 2.

At the n-th stage we have a column C_n of height $h_n = 2^{2n}$ and width $w_n = 1/2^n$ with measure 2^n. We cut the column C_n in half, making two subcolumns of width $w_{n+1} = 1/2^{n+1}$. Then a block of 2^{2n+1} spacers is placed either over the left or over the right subcolumn depending as above on the value of ε_n. As before the right-side subcolumn is placed over the left-side subcolumn giving a column C_{n+1} of height $h_{n+1} = 2^{2n+2}$.

The transformation $T = T_{\bar{\varepsilon}}$ is defined on the set $X = \bigcup_{k=0}^\infty C_k$ as the transformation which moves up the columns linearly. The measure μ denotes the usual Lebesgue measure on the unit interval and moved linearly onto each level of the columns.

It is clear that for various choices of the $\bar{\varepsilon}$'s the map $T_{\bar{\varepsilon}}$ is an infinite ergodic rank-one transformation. The proof that X_0 is an *eww* set with the sequence $\{n_i\}$ (finite sums of odd powers of 2) is a simple modification of the corresponding result for the First Basic Example. We sketch below the construction of $T_{\bar{\varepsilon}}$ and make a few elementary observations describing the first few stages of the construction in order to clarify the properties of the transformation.

We fix $\bar{\varepsilon}$. For notational simplicity we will sometimes suppress the subscript $\bar{\varepsilon}$ and similarly suppress the $\bar{\varepsilon}$ in the columns $C_{n,\bar{\varepsilon}}$ and the transformations $T_{n,\bar{\varepsilon}}$ and $T_{\bar{\varepsilon}}$.

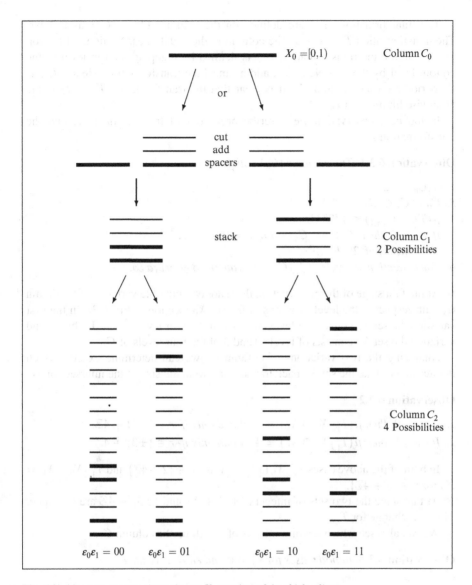

$X_0 = [0,1)$ Column C_0

or

cut
add
spacers

stack   Column C_1
2 Possibilities

Column C_2
4 Possibilities

$\varepsilon_0\varepsilon_1 = 00$ $\varepsilon_0\varepsilon_1 = 01$ $\varepsilon_0\varepsilon_1 = 10$ $\varepsilon_0\varepsilon_1 = 11$

Fig. 6.3 Alternate constructions: the set X_0 consists of the *thicker lines*

The transformations $T_{n,\bar{\varepsilon}}$ are defined on the columns $C_n = C_{n,\bar{\varepsilon}}$ as follows. The transformation $T_{n,\bar{\varepsilon}}$ goes up the column in the usual linear fashion. At the top of the column the transformation $T_{n,\bar{\varepsilon}}$ is defined by mapping the top level to the bottom level by the von Neumann adding machine transformation. Hence $T_{n,\bar{\varepsilon}}$ is set-periodic on the column C_n. It is clear that the transformation $T = T_{\bar{\varepsilon}}$ is the pointwise limit of $\lim T_{n,\bar{\varepsilon}}$.

In the next observation we describe properties of the column $\{C_n\}$ and the transformation T.

Observation 6.3.1 *The columns $\{C_n\}$ satisfy*

1. $\mu(C_n) = 2^n$.
2. $C_0 \subset C_1 \subset C_2 \cdots \nearrow X$.
3. $\mu(T C_n \backslash C_{n+1}) \leq 1/2^{n+1}$.
4. *If* $\varepsilon_n = 0$ *then* $T^j C_n \subset C_{n+1}$ *for* $j = 0, 1, \ldots, 2^{2n+1}$.
5. X_0 *consists of* 2^n *levels of* C_n.
6. *The collection of levels of* $\bigcup C_n$ *generate the σ-algebra $\mathcal{B}$.*

At the first stage of the construction there are two possible variations for column C_1 with respect to the level locations of the set X_0 (see top of Fig. 6.3). In the first variation the set X_0 consists of levels 0 and 1 of the four levels of C_1. In the second variation the set X_0 consists of levels 0 and 3 of the four levels of C_1.

Analyzing the two variations of column C_1 we can determine exactly which iterates of $T_{1,\bar{\varepsilon}}$ take the set X_0 back to itself and what the size of the intersection is.

Observation 6.3.2   .

1. *If* $\varepsilon_0 = 0$, *then* $\mu(T_{1,\bar{\varepsilon}}^r X_0 \cap X_0) = \frac{1}{2}$ *if and only if* $r \in \{\pm 1\} + 4\mathbb{Z}$,
2. *If* $\varepsilon_0 = 1$, *then* $\mu(T_{1,\bar{\varepsilon}}^r X_0 \cap X_0) = \frac{1}{2}$ *if and only if* $r \in \{\pm 3\} + 4\mathbb{Z}$.

In both of the above cases $T_{1,\bar{\varepsilon}}^r X_0 \cap X_0 = \emptyset$ for $r \in \{2 + 4\mathbb{Z}\}$ and $T_{1,\bar{\varepsilon}}^r X_0 \cap X_0 = X_0$ for $r \in \{0 + 4\mathbb{Z}\}$.

At this stage the two sets of integers $\{\pm 1\} + 4\mathbb{Z}$ and $\{\pm 3\} + 4\mathbb{Z}$ are the same. This will change for $T_{2,\bar{\varepsilon}}$.

We can also see which disjoint images of X_0 fill up the column C_1.

Observation 6.3.3 *In both cases for $\varepsilon_0 = 0$ and $\varepsilon_0 = 1$ we have*

$$C_1 = X_0 \,\dot{\cup}\, T_{1,\bar{\varepsilon}}^2 X_0 \ (disj).$$

At the second stage of the construction there are four possible variations for column C_2. These are illustrated at the bottom of Fig. 6.3. The base set X_0 consists of four levels in each variation. In all four cases we can see for which iterates of $T_{2,\bar{\varepsilon}}$ the set X_0 intersects itself in half.

Observation 6.3.4

1. *If* $\varepsilon_0 \varepsilon_1 = 00$, *then* $\mu(T_{2,\bar{\varepsilon}}^r X_0 \cap X_0) = \frac{1}{2}$ *if and only if* $r \in \{\pm 1, \pm 4\} + 16\mathbb{Z}$.

2. *If $\varepsilon_0\varepsilon_1 = 01$, then $\mu(T_{2,\bar\varepsilon}^r X_0 \cap X_0) = \frac{1}{2}$ if and only if $r \in \{\pm 1, \pm 12\} + 16\mathbb{Z}$.*
3. *If $\varepsilon_0\varepsilon_1 = 10$, then $\mu(T_{2,\bar\varepsilon}^r X_0 \cap X_0) = \frac{1}{2}$ if and only if $r \in \{\pm 3, \pm 4\} + 16\mathbb{Z}$.*
4. *If $\varepsilon_0\varepsilon_1 = 11$, then $\mu(T_{2,\bar\varepsilon}^r X_0 \cap X_0) = \frac{1}{2}$ if and only if $r \in \{\pm 3, \pm 12\} + 16\mathbb{Z}$.*

At this second stage we can also determine for which iterates of $T_{2,\bar\varepsilon}$ the intersection is one-fourth. In particular, if for two different values of r, say for $r_1 \neq r_2$, if $\mu(T_{2,\bar\varepsilon}^r X_0 \cap X_0) = \frac{1}{2}$ then for the value $r = r_1 + r_2$ the intersection is $\frac{1}{4}$.

This is easily seen because the larger $|r|$ takes the left subcolumn onto the right subcolumn. Inside the right subcolumn there are two copies of the previous column C_1, and the lower of which is mapped onto the other by the smaller $|r|$.

Observation 6.3.5

1. *If $\varepsilon_0\varepsilon_1 = 00$, then $\mu(T_{2,\bar\varepsilon}^r X_0 \cap X_0) = \frac{1}{4}$ if and only if $r \in \{\pm 1 + \pm 4\} + 16\mathbb{Z}$.*
2. *If $\varepsilon_0\varepsilon_1 = 01$, then $\mu(T_{2,\bar\varepsilon}^r X_0 \cap X_0) = \frac{1}{4}$ if and only if $r \in \{\pm 1 + \pm 12\} + 16\mathbb{Z}$.*
3. *If $\varepsilon_0\varepsilon_1 = 10$, then $\mu(T_{2,\bar\varepsilon}^r X_0 \cap X_0) = \frac{1}{4}$ if and only if $r \in \{\pm 3 + \pm 4\} + 16\mathbb{Z}$.*
4. *If $\varepsilon_0\varepsilon_1 = 11$, then $\mu(T_{2,\bar\varepsilon}^r X_0 \cap X_0) = \frac{1}{4}$ if and only if $r \in \{\pm 3 + \pm 12\} + 16\mathbb{Z}$.*

We also have $T_{2,\bar\varepsilon}^r X_0 \cap X_0 = X_0$ for $r \in \{0 \pm 16\mathbb{Z}\}$ and in all other cases the intersection is empty. We summarize this as:

Observation 6.3.6 *For any $r \notin \{0\} \pm 16\mathbb{Z}$ and not included in the cases of the previous two observations we have $\mu(T_{2,\bar\varepsilon}^r X_0 \cap X_0) = 0$.*

We also note:

Observation 6.3.7 *Suppose $\mu(T_{n,\bar\varepsilon}^r X_0 \cap X_0) = \frac{1}{2^k}$ for $0 \le r < 4^n$, then*

1. $\mu(T_{n+1,\bar\varepsilon}^r X_0 \cap X_0) = \frac{1}{2^k}$,
2. $\mu(T_{n+1,\bar\varepsilon}^{r+4^n+\varepsilon_n 2^{2n+1}}(X_0) \cap X_0) = \frac{1}{2^{k+1}}$.

As before, we can also determine exactly which images of X_0 disjointly fill the column.

Observation 6.3.8 *In all four cases $\varepsilon_0\varepsilon_1 \in \{00, 01, 10, 11\}$*

$$C_2 = X_0 \,\dot\cup\, T_{2,\bar\varepsilon}^2 X_0 \,\dot\cup\, T_{2,\bar\varepsilon}^8 X_0 \,\dot\cup\, T_{2,\bar\varepsilon}^{10} X_0 \ (disj).$$

6.3.2 The Recurrent Sequences for $T_{\bar\varepsilon}$

We now show how to determine completely all the recurrent sequences for the transformation $T = T_{\bar\varepsilon}$. This will match the construction in Sect. 4.1.2. In that section the recurrent sequences were built up from the sequences N_k and the sequences N_k were all built up from N_1. With that in mind we make the following definition.

For a fixed $\bar{\varepsilon}$ let us define the *foundational sequence* as

$$\mathbb{F} = \mathbb{F}(\bar{\varepsilon}) = \{f_i = 2^{2i} + \varepsilon_i 2^{2i+1} : i = 0, 1, \ldots\}.$$

Proposition 6.3.9. *For all $f \in \mathbb{F}$ we have:*

$$\mu(T_{\bar{\varepsilon}}^{\pm f} X_0 \cap X_0) = \frac{1}{2}$$

and $\mathbb{F}$ is a recurrent sequence for $T_{\bar{\varepsilon}}$.

Proof. The proof follows from the construction of the transformation $T_{\bar{\varepsilon}}$. We notice that at the n-th stage when the column C_n of height 4^n is cut in half there are two subcolumns each containing half of the set X_0. When these are stacked, with the spacers added in the appropriate location, we have the lower half mapping onto the upper half under the iterate T^r with $r = 4^n + \varepsilon_n 2^{2n+1}$. The upper half is simultaneously mapped into spacers. By Proposition 3.3.2 then $\mathbb{F}$ is a recurrent sequence for $T_{\bar{\varepsilon}}$.   □

The definition of $\mathbb{F}$ matches the definition of N_1 in Eq. (4.7) for the First Basic Example and so we make the following definitions.

Let us set $N_0 = \{0\}$ and

$$N_k = N_{k,\bar{\varepsilon}} = \{n : n = \sum_{i=1}^{k} \pm f_{p_i}, \ f_p \in \mathbb{F}_{\bar{\varepsilon}}, \ 0 < p_1 < p_2 < \cdots < p_k\}.$$

Each $N_{k,\bar{\varepsilon}}$ is a recurrent sequence for $T_{\bar{\varepsilon}}$ and the sets $N_{k,\bar{\varepsilon}}, N_{l,\bar{\varepsilon}}$ are disjoint when $k \neq l$. It follows that we have the analog to Eq. (4.8) for the First Basic Example. That is, if we take a finite sum $r = f_{p_1} + f_{p_2} + \cdots + f_{p_k}$ with $f_{p_i} \in \mathbb{F}$ and $0 < p_1 < p_2 < \cdots < p_k$ we can determine the size of $T^r X_0 \cap X_0$. We first look at $T^{p_k} X_0$ on column C_{p_k}. As usual the left subcolumn maps into the right subcolumn giving an intersection of measure $1/2$. Inside this subcolumn there are multiple copies of column $C_{p_{k-1}}$. Each of these has its relative left subcolumns mapping onto the relative right subcolumns resulting in an intersection of measure $1/4$. Continuing with this analysis we conclude with the following observation.

Observation 6.3.10 $\mu(T^n X_0 \cap X_0) = \begin{cases} 1/2^k & \text{if } n \in N_k, \\ 0 & \text{otherwise.} \end{cases}$

New recurrent sequences can be constructed from the N_k. We can shift them, take finite unions and then take infinite subsets. It follows that all recurrent sequences for a fixed transformation $T_{\bar{\varepsilon}}$ can be obtained this way.

Next, suppose $\mathbb{R}$ is a recurrent sequence. Then there is a measurable set A for which $\liminf \mu(T^{r_i} A \cap A) = 4\delta > 0$. From the construction we can find a column

C_k such that $\mu(A \backslash C_k) < \delta$. Hence $\liminf \mu(T^{r_i} C_k \cap C_k) > \delta > 0$. Recalling that W consists of a finite union of some levels of C_k we can shift up and down and find an $l > k$ and integers n_i, m_i, so that

$$\mu(T^{r_i + n_i - m_i} X_0 \cap X_0) \geq 1/2^l \text{ for } 0 \leq n_i, m_i < 2^{2k}.$$

This means that $r_i + n_i - m_i \in \bigcup_{j=0}^{l} N_j$.

Using the above the following proposition gives a complete description of the recurrent sequences for the transformation $T_{\bar{\varepsilon}}$. The proof is identical to the proof of Theorem 4.1.1.

Proposition 6.3.11. *A sequence* $\mathbb{R} = \{r_i : i = 0, 1, \ldots\}$ *is a recurrent sequence for the transformation* $T = T_{\bar{\varepsilon}}$ *if and only if there exist two positive integers* l *and* s *such that for* $N_k = N_{k,\bar{\varepsilon}}$

$$\mathbb{R} \subset \bigcup_{n=-s}^{s} \bigcup_{k=0}^{l} (N_k + n).$$

If $\bar{\varepsilon}$ and $\bar{\eta}$ differ in only a finite number of coordinates then there exist $s > 0$ such that $\mathbb{F}_{\bar{\varepsilon}} \subset \bigcup_{n=-s}^{s} (\mathbb{F}_{\bar{\eta}} + n)$ and the corresponding families of recurrent sequences are the same. However, if $\bar{\varepsilon}$ and $\bar{\eta}$ differ in an infinite number of coordinates then $\mathbb{F}_{\bar{\varepsilon}}$ cannot be a recurrent sequence for $T_{\bar{\eta}}$.

As a final remark we point out that when $\bar{\varepsilon}$ and $\bar{\eta}$ differ in a finite number of coordinates the two transformations are isomorphic; however, the isomorphism does not map the relative X_0's onto each other.

The next theorem summarizes the previous discussion.

Theorem 6.3.12. *For every choice of* $\bar{\varepsilon}$ *the transformation* $T_{\bar{\varepsilon}}$ *is an infinite ergodic transformation of* α-*type with* $\alpha = 1/2$. *The set* X_0 *(the non-dyadic rationals in* $[0, 1)$) *is an eww set with the sequence* $\{n_i\}$ *as given in (4.4) of Chap. 4. In addition, the two transformations* $T_{\bar{\varepsilon}}$ *and* $T_{\bar{\eta}}$ *are not isomorphic if and only if* $\bar{\varepsilon}$ *and* $\bar{\eta}$ *differ in an infinite number of coordinates.*

6.4 Growth Distributions for a Transformation

Let T be an infinite ergodic transformation defined on $(X, \mathscr{B}, \mu)$ and G a set of positive finite measure. Proposition 2.2.1 implies that the union of the successive images of the set G under powers of the transformation T grows in size to fill up the whole space X. In particular, this says $\lim_{i \to \infty} \mu(\bigcup_{j=0}^{i-1} T^j G) = \infty$. On the other hand, this growth cannot be too fast and must be slow enough to satisfy the ergodicity of the transformation T: for instance we must have

$\lim_{i\to\infty} \frac{1}{i}\mu(\bigcup_{j=0}^{i-1} T^j G) = 0$. However, there also exists a lower bound to the rate of this growth that is determined by the ww sequences of the transformation. In fact the lower bound is related to the "thickness" of the ww sequences of T. We clarify the above in Theorem 6.4.2 that follows. First we introduce some notation.

Let T be an infinite ergodic transformation defined on $(X, \mathcal{B}, \mu)$, and let G be a set of finite positive measure. We consider the following sequence of numbers:

$$\mathbf{g} = (g(1), g(2), \ldots) \quad \text{where} \quad g(i) = \mu\left(\bigcup_{l=0}^{i-1} T^l G\right).$$

(Aaronson [1] calls this the wandering rate for T.) We note that the sequence $\mathbf{g} = (g(i) : i = 1, 2, \cdots)$ measures the growth of the first i floors of the skyscraper over the set G. Since T is ergodic, this implies that $g(i)$ increases to ∞, and $g(i) - g(i-1)$ decreases to 0 (see also [21]). We use the following terminology.

Definition 6.4.1. Let T be an infinite ergodic transformation defined on $(X, \mathcal{B}, \mu)$. An infinite sequence of real numbers $\mathbf{g} = (g(1), g(2), \ldots)$ is a *growth distribution* for T if there is a set $G \in \mathcal{B}$ such that $g(i) = \mu\left(\bigcup_{l=0}^{i-1} T^l G\right)$ for $i \geq 1$.

In the following theorem we show a few properties of growth distributions for an infinite ergodic transformation (see [30, 44]).

Theorem 6.4.2. *Let T be an infinite ergodic transformation defined on $(X, \mathcal{B}, \mu)$. Let $\mathbf{g} = (g(1), g(2), \ldots)$ be a growth distribution for T and let $\{n_i : i \geq 1\}$ be a ww sequence for T. Then*

1. $\lim_{i\to\infty} g(i) = \infty$,

2. $\lim_{i\to\infty} \dfrac{g(i)}{i} = 0$, *and*

3. $\lim_{i\to\infty} \dfrac{g(n_i)}{i} > 0$.

Proof. Let G be a set of positive measure such that $g(i) = \mu(\bigcup_{l=0}^{i-1} T^l G)$ for $i \geq 1$. As mentioned above, T ergodic implies that the first i floors of the skyscraper $\bigcup_{l=0}^{i-1} T^l G$ increase to X, and the measure of the i-th floor of the skyscraper over the set G tends to 0. This proves properties 1 and 2.

Let A be a ww set for the ww sequence $\{n_i\}$. Since T is ergodic, there is a positive integer k so that $B = T^k A \cap T^{-1} G$ has positive measure. Therefore,

$$g(n_i) = \mu\left(\bigcup_{l=0}^{n_i-1} T^l G\right) \geq \mu\left(\bigcup_{l=1}^{i} T^{n_l} T^{-1} G\right) \geq \mu\left(\bigcup_{l=1}^{i} T^{n_l} B\right) = \sum_{l=1}^{i} \mu(T^{n_l} B) = i\mu(B).$$

Dividing both sides of the above inequality by i completes the proof.   □

It is clear that growth distributions for an infinite ergodic transformation are an isomorphism invariant. In [44] Kakutani indicates how to use Theorem 6.4.2 to construct a sequence of transformations T_n defined on $(X, \mathscr{B}, \mu)$ no two of which are isomorphic. We show this in what follows (see [18]).

Theorem 6.4.3. *Let* $\tilde{\mathbf{g}} = (\tilde{g}(1), \tilde{g}(2), \ldots)$ *be a growth distribution for some infinite ergodic transformation. Then there exists an infinite ergodic transformation T built as a skyscraper over an odometer for which $\tilde{\mathbf{g}}$ is not a growth distribution for T.*

Proof. Below we construct an infinite ergodic transformation T defined on $(X, \mathscr{B}, \mu)$, the real line with Lebesgue measure, which has a growth distribution $\mathbf{g}$ satisfying the following two conditions:

1. T admits a *ww* sequence $\{n_i : i \geq 1\}$ satisfying $1 \leq \dfrac{g(n_i)}{i} \leq 2$ for all $i \geq 1$, and

2. $\lim\limits_{i \to \infty} \dfrac{\tilde{g}(i)}{g(i)} = 0.$

These two conditions imply $\lim_{i \to \infty} \tilde{g}(n_i)/i = 0$. Thus, by Theorem 6.4.2, the distribution $\tilde{\mathbf{g}}$ cannot be a growth distribution for the transformation T. We complete the proof by constructing the example that follows. $\qquad\square$

Example: The transformation T is a skyscraper over the dyadic odometer S on X_0, the unit interval with the dyadic rationals removed. It will be a generalization of the First Basic Example of Chap. 4. We exhibit an *eww* sequence $\{n_i\}$ for T with X_0 as the *eww* set. If we let $g(i) = \mu\left(\bigcup_{l=1}^{i} T^{l-1}(X_0)\right)$ be the growth distribution of the skyscraper for T over X_0, then for all $i \geq 1$

$$1 \leq \frac{g(n_i)}{i} \leq 2. \tag{6.7}$$

Furthermore, for the given growth distribution $\tilde{\mathbf{g}}$, we can choose parameters in the construction of T, so that the resulting transformation T also satisfies

$$\lim_{i \to \infty} \frac{\tilde{g}(i)}{g(i)} = 0, \tag{6.8}$$

and thus $\tilde{\mathbf{g}}$ cannot be a growth distribution for that T.

Similar to the First Basic Example of Chap. 4 we let X_0 be the unit interval with the dyadic rationals removed. For $k = 0, 1, 2, \ldots$ we denote the dyadic interval of length $1/2^{k+1}$ by $A_k = \{x \in X_0 : 1 - \frac{1}{2^k} < x < 1 - \frac{1}{2^{k+1}}\}$. On X_0 we consider the dyadic odometer transformation S, and let the infinite ergodic transformation T be a skyscraper over the dyadic odometer S on X_0. For each $k \geq 0$ the interval $A_k \subset X_0$ will form the base of the column of height h_k for T. The h_k's, an increasing sequence of integers which we specify later, will depend on a given sequence of positive integers $p_k > 0$. The p_k's are parameters in the construction that we can select at our discretion. We will see that for any choice of the $p'_k s$ condition (6.7) holds. Incidentally, if we choose $p_k = 1$ for all k then the resulting transformation will be the First Basic Example of Chap. 4. Initially, we do the construction ignoring condition (6.8). At the end of the construction we observe that by choosing p_k sufficiently large at each stage, we can guarantee that (6.8) is satisfied. In addition to specifying the h_k's, we describe an increasing sequence of integers n_k so that the sets $T^{n_k} X_0$ partition the space X. This exhibits an explicit *eww* set and sequence for the transformation T. Furthermore, for $g(i) = \mu(\bigcup_{l=1}^{i} T^{l-1} X_0)$ we will show that $i \le g(n_i) \le 2i$ for all $i \ge 1$.

The measure space $(X, \mathscr{B}, \mu)$ and transformation T are defined as the skyscraper over X_0, with the action on the base X_0 given by S, the odometer. Thus, the space X is the disjoint union

$$X = \bigcup_{k=0}^{\infty} \bigcup_{l=1}^{h_k} T^{l-1} A_k \, (disj),$$

with $\mathscr{B}$ and μ representing the Lebesgue measurable sets and Lebesgue measure respectively. We keep track of the growth of the skyscraper by means of the following sets:

$$B_0 = \bigcup_{l=1}^{h_0} T^{l-1} X_0 \, , \qquad G_0 = B_0,$$

and for $k = 1, 2, \ldots$

$$B_k = \bigcup_{j=k}^{\infty} \bigcup_{l=h_{k-1}+1}^{h_k} T^{l-1} A_j \, , \qquad G_k = G_{k-1} \cup B_k.$$

We note that G_k is the union of the first h_k floors of the skyscraper (its measure is the growth $g(h_k)$), and B_k is the block of "new floors" which we add to the previous portion of the skyscraper G_{k-1}. We elaborate on this, making precise our choice of h_k and determining the appropriate n_k. At each stage of the construction we will define the sequence of integers n_i so that (6.7) holds, and we will keep track of the indices i for which n_i is defined by means of an auxiliary sequence q_k (depending on the parameter sequence p_k).

Figure 6.4 below is important in understanding the details of this example. In the description which follows we also give a parallel cutting and stacking description of the transformation T. However, the skyscraper description given in Fig. 6.4 is sufficient.

Let $\{p_k > 0 : p_k \in \mathbf{N}\}$ be any sequence of positive integers. Later on we will need p_k's to be large in order to show that (6.8) holds.

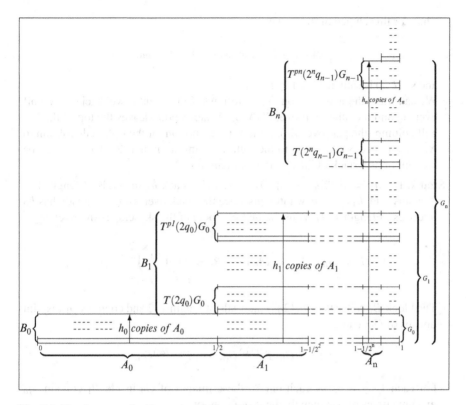

Fig. 6.4 The skyscraper for T over the dyadic odometer

Step 1. Over X_0 we stack p_0 copies of X_0. Thus $A_0 = (0, 1/2)$ will be the base of the column of height $h_0 = p_0$. We set B_0 to be the first h_0 floors of the skyscraper:

$$B_0 = \bigcup_{l=1}^{h_0} \bigcup_{j=0}^{\infty} T^{l-1} A_j, \quad G_0 = B_0.$$

Since $\{T^l X_0 : l = 0, \ldots, p_0 - 1\}$ are all disjoint, these give our first $p_0 - 1$ nonzero terms in the sequence $\{n_i\}$. Thus we set $q_0 = p_0$, and for each i, $0 \leq i < q_0$, we let

$$n_i = i.$$

We observe that $\{T^{n_i} X_0 : i = 0, \ldots, q_0 - 1\}$ are disjoint, and the first h_0 floors of the skyscraper can be written as the disjoint union

$$G_0 = \bigcup_{i=0}^{q_0-1} T^{n_i} X_0 \ (disj).$$

Thus for this choice of n_i we have

$$g(n_i) = i \quad \text{for each} \quad 1 \le i < q_0,$$

and so (6.7) is satisfied for $1 \le i < q_0$.

We can view the action T on G_0 as follows. Cut G_0 into two stacks, one half over A_0 and the other half over $X_0 \setminus A_0$. When a point leaves the top of the left half column (the part over A_0), it maps to the bottom of the right half column to $X_0 \setminus A_0$ (since S moves A_0 to the right by translation of $1/2$). Thus G_0 can be thought of as a stack of $2q_0$ intervals of length $1/2$.

Step 2. Let $h_1 = p_1(2q_0) + h_0$. Over $X_0 \setminus A_0$, stack h_1 intervals of length $1/2$ (we only add $h_1 - h_0$ new intervals since the stack over $X_0 \setminus A_0$ already has h_0 intervals of length $1/2$). Thus the first h_1 floors of the skyscraper are given by

$$G_1 = G_0 \cup B_1, \quad \text{where} \quad B_1 = \bigcup_{l=h_0+1}^{h_1} \bigcup_{j=1}^{\infty} T^{l-1} A_j.$$

The block B_1 consists of $p_1(2q_0)$ intervals of length $1/2$ and contains p_1 disjoint images of G_0; namely

$$T^{(2q_0)} G_0, T^{2(2q_0)} G_0, \cdots, T^{p_1(2q_0)} G_0.$$

For each $1 \le j \le p_1$ each one of these images of G_0 inside B_1 contains q_0 disjoint copies of X_0; namely the disjoint union

$$T^{j(2q_0)} G_0 = \bigcup_{k=0}^{q_0-1} T^{j(2q_0)+n_k} X_0 \ (disj).$$

As j ranges through $0, 1, 2, \ldots, p_1$ it is these $(p_1 + 1)q_0$ disjoint images of X_0 that determine the set of indices n_i (the first q_0 coming from Step 1). Namely, for $q_1 = (p_1 + 1)q_0$ we extend the definition of n_i for $q_0 \le i < q_1$ as follows:

for $i = jq_0 + k$ where $1 \le j \le p_1$ and $0 \le k < q_0$ we set $n_i = j(2q_0) + k$.

We note that when $i = jq_0 + k$ we can write $n_i = j(2q_0) + k = n_{jq_0} + n_k$. Furthermore, since we chose these indices so that $\{T^{n_i} X_0 : i = 0, \ldots, q_1 - 1\}$ are disjoint, we need only verify the growth inequality $i \le g(n_i) \le 2i$ for each i, with $i < q_1$. Since the first n_i floors of the skyscraper over X_0 contain i disjoint copies of X_0, we see that $i \le g(n_i)$ trivially holds for each i in this range.

We note that for each $j = 1, 2, \ldots, p_1$ the first $(2j - 1)q_0$ floors of the skyscraper fill up the j sets in the union $G_0 \cup T^{2q_0} G_0 \cup T^{2(2q_0)} G_0 \cup \cdots \cup T^{(j-1)(2q_0)} G_0$ exactly. Thus, since $n_{jq_0-1} = (2j - 1)q_0 - 1$,

$$\bigcup_{l=1}^{n_{jq_0}-1} T^{l-1} X_0 = \bigcup_{i=0}^{j-1} T^{i(2q_0)} G_0 = \bigcup_{i=0}^{jq_0-1} T^{n_i} X_0 \; (disj).$$

We also note that the next k floors of the skyscraper for any $k = 1, \ldots, q_0 - 1$ are all subsets of $T^{j(2q_0)}(G_0)$. In fact

$$\bigcup_{l=n_{jq_0}}^{n_{(j+1)q_0}-1} T^l X_0 \subset T^{n_{jq_0}} G_0.$$

It follows that for any i, with $jq_0 \le i < (j+1)q_0$, we have $g(n_i) \le jq_0 + q_0$. Using $i \le g(n_i)$ it follows that $jq_0 \le g(n_i) < (j+1)q_0$ for $jq_0 \le i < (j+1)q_0$, and therefore $g(n_i)/i \le (j+1)/j \le 2$. Thus (6.7) is satisfied for all $1 \le i < q_1$. We also note that

$$G_1 = \bigcup_{l=1}^{n_{q_1}-1} T^{l-1} X_0 = \bigcup_{i=0}^{q_1-1} T^{n_i} X_0 \; (disj).$$

The action of T on G_1 is as follows. Since G_1 is a stack of p_1 copies of G_0 stacked over G_0, G_1 is a stack of $(p_1 + 1)2q_0 = 2q_1$ intervals of length $1/2$. Under the action of S on the base this stack is cut in half and the left half stacked under the right half. Thus the set G_1 can also be thought of as a stack of $2^2 q_1$ intervals of length $1/2^2$.

Step s. Proceeding inductively, we assume that the integers h_s, q_s and $\{n_i : 0 \le i < q_s\}$ have been defined so that

$$G_s = \bigcup_{l=1}^{n_{q_s}-1} T^{l-1} X_0 = \bigcup_{i=0}^{q_s-1} T^{n_i} X_0 \; (disj),$$

and (6.7) is satisfied for all $1 \le i < q_s$.

We let $h_{s+1} = p_{s+1}(2^{s+1} q_s) + h_s$. We note that the set B_{s+1} consists of $h_{s+1} - h_s$ intervals of length $1/2^{s+1}$. The set G_s can also be thought of as a stack of $2^{s+1} q_s$ intervals of length $1/2^{s+1}$.

Thus B_{s+1} consists of $p_{s+1}(2^{s+1} q_s)$ intervals of length $1/2^{s+1}$ over G_s and contains p_{s+1} disjoint images of G_s; namely

$$T^{(2^{s+1} q_s)} G_s, \; T^{2(2^{s+1} q_s)} G_s, \ldots, \; T^{p_{s+1}(2^{s+1} q_s)} G_s.$$

Moreover, each one of these images of G_s inside B_{s+1} contains q_s disjoint copies of X_0; namely for $1 \leq j \leq p_{s+1}$

$$T^{j(2^{s+1}q_s)}G_s = \bigcup_{k=0}^{q_s-1} T^{j(2^{s+1}q_s)+n_k} X_0 \ (disj).$$

It is these disjoint images of X_0 under the powers of T that determine the next set of indices n_i. We set $q_{s+1} = (p_{s+1} + 1)(q_s)$ and extend the definition of n_i for $q_s \leq i < q_{s+1}$ as follows. First we write $i = jq_s + k$ where $1 \leq j \leq p_{s+1}$ and $0 \leq k < q_s$. We set

$$n_i = j(2^{s+1}q_s) + n_k (= n_{jq_s} + n_k).$$

As the n_i have been chosen so that $T^{n_i} X_0$ are disjoint for $0 \leq i < q_{s+1}$, we need only check property (6.7) for these n_i. We note that for $j = 1, 2, \ldots, p_{s+1}$

$$\bigcup_{l=1}^{n_{jq_s}-1} T^{l-1} X_0 = \bigcup_{i=0}^{j-1} T^{i(2^{s+1}q_s)}G_s = \bigcup_{i=0}^{jq_s-1} T^{n_i} X_0 \ (disj),$$

and the next n_k floors (for $k < q_s$) of the skyscraper are subsets of $T^{j(2^{s+1}q_s)}G_s$. In fact

$$\bigcup_{l=n_{jq_s}}^{n_{(j+1)q_s}-1} T^l X_0 \subset T^{n_{jq_s}} G_s.$$

Thus for each $j = 1, 2, \ldots, p_{s+1}$ we have $g(n_i) \leq jq_s + q_s = (j + 1)q_s$ for $jq_s \leq i < (j + 1)q_s$. The inequality $i \leq g(n_i)$ is clear, and it follows that $jq_s \leq g(n_i) \leq (j + 1)q_s$ for $jq_s \leq i < (j + 1)q_s$. Thus as in **Step 2**, (6.7) is satisfied for all $1 \leq i < q_{s+1}$.

We note that

$$G_{s+1} = \bigcup_{l=1}^{n_{q_{s+1}}-1} T^{l-1} X_0 = \bigcup_{i=0}^{q_{s+1}-1} T^{n_i} X_0 \ (disj),$$

and can also be thought of as a stack of $2^{s+2}q_{s+1}$ intervals of length $1/2^{s+2}$. When a point leaves the top left of G_{s+1}, it goes to the bottom of the right half of G_{s+1}.

In this way the ergodic measure-preserving transformation T is defined on the infinite measure space $(X, \mathcal{B}, \mu)$ where $X = \bigcup_i G_i$. We have also constructed the ww sequence $\{n_i : i \geq 0\}$ with the set X_0 satisfying

$$1 \leq \frac{g(n_i)}{i} \leq 2 \quad \text{for all } i \geq 1.$$

All that remains to complete this example (and therefore complete the proof of Theorem 6.4.3) is to show that by choosing the p_k's large enough we can ensure that (6.8) also holds. Namely,

$$\lim_{i \to \infty} \frac{\tilde{g}(i)}{g(i)} = 0,$$

where $g(i) = \bigcup_{l=1}^{i} T^{l-1} X_0$ and $\tilde{g}$ is the given growth distribution.

Without loss of generality we will assume that the growth distribution $\tilde{g}$ has $\tilde{g}(1) = 1$. We denote by $\tilde{f}(i) = \tilde{g}(i) - \tilde{g}(i-1)$ the measure of the ith floor of the growth distribution $\tilde{g}$ and note the floor distribution has $\tilde{f}(i)$ decreasing to 0. At each stage in the previous construction we start with $p_0 = h_0$ large enough so that $1 = f(i) = f(h_0) = \mu(T^{h_0-1}(X_0)) \geq \sqrt{\tilde{f}(h_0)}$ for all $i \leq h_0$. Then for each $k \geq 1$ we choose p_k sufficiently large, and therefore h_k will be sufficiently large also, so that

$$\frac{1}{2^{k+1}} \geq \sqrt{\tilde{f}(h_k)}.$$

By the choice of p_k's for i's in the range $h_k \leq i \leq h_{k+1}$ we have

$$\frac{1}{2^{k+1}} = f(i) \geq \sqrt{\tilde{f}(h_k)} \geq \sqrt{\tilde{f}(i)}.$$

Next we let k be any (large) integer, and for $n \geq h_k$ we consider the ratio

$$\begin{aligned}
\frac{g(n)}{\tilde{g}(n)} &= \frac{\sum_{i=1}^{n} f(i)}{\sum_{i=1}^{n} \tilde{f}(i)} \\[1em]
&\geq \frac{\sum_{i=h_k}^{n} f(i)}{\sum_{i=1}^{n} \tilde{f}(i)} \\[1em]
&\geq \frac{\sum_{i=h_k}^{n} \frac{\tilde{f}(i)}{\sqrt{\tilde{f}(i)}}}{\sum_{i=1}^{n} \tilde{f}(i)} \\[1em]
&\geq \left(\frac{\sum_{i=h_k}^{n} \tilde{f}(i)}{\sum_{i=1}^{n} \tilde{f}(i)} \right) 2^{k+1}.
\end{aligned}$$

Letting $n \to \infty$ gives $\liminf(g(n)/\tilde{g}(n)) \geq 2^{k+1}$. Since k was arbitrary we have $\lim_{n \to \infty}(g(n)/\tilde{g}(n)) = \infty$. Thus property (6.8) is also satisfied. This shows that the example does not admit the growth distribution $\tilde{\mathbf{g}}$.   □

J. Aaronson has shown us (as outlined in [18]) how one can use an irreducible, null-recurrent Markov chain (instead of a skyscraper over an odometer) not admitting a given growth distribution $\tilde{\mathbf{g}}$.

Chapter 7
Integer Tilings

In this chapter we study infinite tilings of the integers and we explore the connection between infinite tilings of the integers and infinite ergodic theory. This will reveal a structure for certain *eww* sequences. With this structure we will extend the examples of Chap. 4.

7.1 Infinite Tilings of the Integers

Every infinite ergodic transformation has at least one (and in general more than one) *eww* sequence. These *eww* sequences are tied tightly to the transformation in the same way the period of a set periodic transformation is tied to the transformation. The collection of *eww* sequences is an isomorphism invariant.

This leads to the question of which sequences of integers can be *eww* for some transformation. A starting point is the observation (Theorem 7.2.1) that every *eww* sequence is a member of a pair of infinite direct summands of the integers (defined below). A goal in this chapter is to study this connection between infinite ergodic theory and infinite complementing pairs of subsets for the integers. The purpose is twofold. Using infinite ergodic theory we can discover new results in the study of complementing pairs for the integers, and using the theory of additive combinatorics we hope to discover new insights in infinite ergodic theory. We first define what it means for a subset of integers $\mathbb{A}$ to tile the integers $\mathbb{Z}$.

Definition 7.1.1.

- Given two subsets $\mathbb{A}$ and $\mathbb{B}$ of the integers $\mathbb{Z}$ we write the *sumset* as
  $\mathbb{A} + \mathbb{B} = \{n \in \mathbb{Z} : n = a + b \text{ for } a \in \mathbb{A} \text{ and } b \in \mathbb{B}\}$.
- When every element in the sumset $\mathbb{A} + \mathbb{B}$ can be written in only one way as a sum of elements from $\mathbb{A}$ and $\mathbb{B}$ we say the sumset of the pair $\mathbb{A}$ and $\mathbb{B}$ is *direct*, and we write $\mathbb{A} \oplus \mathbb{B}$.

© Springer Japan 2014
S. Eigen et al., *Weakly Wandering Sequences in Ergodic Theory*,
Springer Monographs in Mathematics, DOI 10.1007/978-4-431-55108-9_7

- A set of integers $\mathbb{A} \subset \mathbb{Z}$ is said to *tile the integers* if there exists a set $\mathbb{B} \subset \mathbb{Z}$ so that $\mathbb{A} \oplus \mathbb{B} = \mathbb{Z}$. We also say that $\mathbb{A}$ and $\mathbb{B}$ are a *complementing pair in* $\mathbb{Z}$ or a *direct sumset* for $\mathbb{Z}$. Often we shall use the terminology: a set $\mathbb{B}$ is a complementing set (in $\mathbb{Z}$) of $\mathbb{A}$ to mean $\mathbb{A}$ and $\mathbb{B}$ is a complementing pair in $\mathbb{Z}$, or $\mathbb{A} \oplus \mathbb{B} = \mathbb{Z}$.

The definition allows that one of the pair $\mathbb{A}, \mathbb{B}$ may be finite. When one of the summands in the pair is finite a great deal is known [4]. We summarize some of these results in the next subsection. For the tilings of $\mathbb{Z}$ that arise from infinite ergodic transformations both summands will be infinite, and in general none of the tools of the finite tile case can be applied. On the other hand, using infinite ergodic theory leads to results dramatically different from the finite tile case. By an *infinite tiling of* $\mathbb{Z}$ we will always mean both $\mathbb{A}$ and $\mathbb{B}$ are infinite. In general we will assume that the pair is normalized so that $\mathbb{A} \cap \mathbb{B} = \{0\}$.

7.1.1 Structure of Complementing Pairs in $\mathbb{N}$

De Bruijn [6] raised the question of the structure of complementing pairs $(\mathbb{A}, \mathbb{B})$ for $\mathbb{Z}$. He already knew a structure theorem for the nonnegative integers $\mathbb{N}$ (see Theorem 7.1.2). The latter structure leads directly to a construction of a family of transformations similar to the First Basic Example of Chap. 4 (see Sect. 7.7). For complementing pairs for $\mathbb{Z}$ the case that one of $\mathbb{A}$ or $\mathbb{B}$ is finite leads to significant results, though there are still many open problems. In Sect. 7.1.3 two results are presented which show that in general there is no structure or effective characterization for all pairs of infinite complementing sets in $\mathbb{Z}$. Despite this, and some further examples, we show that results can still be obtained for a large class of infinite complementing pairs in $\mathbb{Z}$. The tools we use come from infinite ergodic theory and p-adic analysis as it arises in infinite ergodic theory. Specifically, we analyze the structure of *eww* sequences for various transformations.

The structure of complementing pairs for the nonnegative integers $\mathbb{A} \oplus \mathbb{B} = \mathbb{N}$ in the next theorem appears implicitly in de Bruijn [7], and it was rediscovered by Vaidya [53] in 1966. We explain at the end of this chapter how the structure theorem for complementing pairs for the nonnegative integers leads directly to a class of infinite ergodic transformations, one of which is isomorphic to the First Basic Example (see Sect. 7.7).

Theorem 7.1.2. *Two infinite subsets of the nonnegative integers* $\mathbb{A}$ *and* $\mathbb{B}$ *are a complementing pair in* $\mathbb{N}$ *if and only if there exists an infinite sequence of integers* $\mathcal{M} = \{m_i : i \geq 1\}$ *with* $m_i \geq 2$ *for all* i, *such that every* $a \in \mathbb{A}$ *and* $b \in \mathbb{B}$ *is a finite sum of the following form:*

$$a = \sum_{i=0}^{\infty} c_{2i} M_{2i} \quad and \quad b = \sum_{i=0}^{\infty} c_{2i+1} M_{2i+1}$$

where the coefficients c_i satisfy $0 \leq c_i < m_{i+1}$, and $M_0 = 1$, $M_i = \prod_{j=1}^{i} m_j$, $i \geq 1$.

We note that when one of the pair $\mathbb{A}, \mathbb{B}$ of subsets of $\mathbb{N}$ is finite, they still form a complementing pair in $\mathbb{N}$ but in this case $\mathcal{M} = \{m_1, \ldots, m_r\}$ is finite and the final coefficient is unbounded; i.e. $0 \leq c_r < \infty$.

Theorem 7.1.2 shows that if a set $\mathbb{A}$ has a complementing set in $\mathbb{N}$ then that complementing set is unique. This contrasts with complementing sets in $\mathbb{Z}$ where we will show that *eww* sequences have a continuum of complementing sets in $\mathbb{Z}$.

Long [48] gave a similar characterization for complementing pairs in $\mathbb{N}_n = \{0, 1, \ldots, n - 1\}$. He showed that the number $C(n)$ of complementing pairs in $\mathbb{N}_n$ is the same as the number of ordered nontrivial factorizations of n. He further showed that $2C(n) = \sum_{d|n} C(d)$. Extensions of Theorem 7.1.2 to $\mathbb{N} \times \mathbb{N}$ were given by Hansen [36] and Niven [50].

We will also see that there is a strong connection between the infinite complementing pairs in $\mathbb{N}$ and infinite ergodic theory. In particular, given the two sequences $\mathbb{A}$ and $\mathbb{B}$ derived from the sequence $\mathcal{M}$, a pair of ergodic transformations (one for $\mathcal{A}$ and the other for $\mathcal{B}$) can be constructed. The simplest case with $m_i = 2$ for all i gives the First Basic Example and a dual transformation. We will show how properties of the sequence $\mathcal{M}$ are reflected in properties of the associated transformations.

7.1.2 Complementing Pairs in $\mathbb{Z}$ When $\mathbb{A}$ or $\mathbb{B}$ Is Finite

For complementing pairs in $\mathbb{Z}$ a great deal is known when one of the pair $\mathbb{A}, \mathbb{B}$ is finite (without loss of generality we will assume $\mathbb{A}$ is the finite set of the pair). Newman [49] found necessary and sufficient conditions for a finite set $\mathbb{A}$ of prime power size to tile the integers. Coven and Meyerowitz [4] have a structure theorem for a finite set $\mathbb{A}$ when $|\mathbb{A}|$ has at most two prime factors.

Assume $\mathbb{A}$ is finite and $\mathbb{A} \oplus \mathbb{B} = \mathbb{Z}$. Then $\mathbb{B}$ must be periodic [7, 34]. That is, $\mathbb{B} = \mathbb{B}' \oplus N\mathbb{Z}$, where $\mathbb{B}'$ is a finite set of integers and $|\mathbb{A}| \cdot |\mathbb{B}'| = N$. In this case we have $\mathbb{A} \oplus \mathbb{B}' = \mathbb{Z}/N\mathbb{Z}$.

This allows one to analyze the set $\mathbb{A}$ using the theory of direct sum factorizations of finite cyclic groups, character theory and cyclotomic polynomials.

It is clear that both $\mathbb{A}$ and $\mathbb{B}'$ can be shifted without losing $\mathbb{A} \oplus \mathbb{B}' = \mathbb{Z}/N\mathbb{Z}$. Hence we can assume $\mathbb{A} = \{0 = a_1 < a_2 < \cdots < a_n\}$ and $\mathbb{B}' = \{0 = b_1 < b_2 < \cdots < b_m\}$. We define the polynomials $p_{\mathbb{A}}(x) = \sum_{a \in \mathbb{A}} x^a$ and $p_{\mathbb{B}'}(x) = \sum_{b \in \mathbb{B}'} x^b$. Then $p_{\mathbb{A}}(x) \cdot p_{\mathbb{B}'}(x) = \sum_{0}^{N-1} x^i \bmod (x^N - 1)$. Let $\phi_s(x)$ be the s-th cyclotomic polynomial. For every s which divides N, $\phi_s(x)$ divides $\sum_{0}^{N-1} x^i$ and so must divide $p_{\mathbb{A}}(x)$ or $p_{\mathbb{B}'}(x)$. If $s = p^\alpha$, a prime power, then $\phi_s(x)$ will divide only one of $p_{\mathbb{A}}(x)$ and $p_{\mathbb{B}'}(x)$. If s is a product of powers of different primes then it is possible for $\phi_s(x)$ to divide both $p_{\mathbb{A}}(x)$ and $p_{\mathbb{B}'}(x)$.

We define $S_{\mathbb{A}}$ as the set of prime powers s such that the s-th cyclotomic polynomial $\phi_s(x)$ divides $p_{\mathbb{A}}(x)$. Coven and Meyerowitz [4] consider the following two conditions:

(T1) $p_{\mathbb{A}}(1) = \prod_{s \in S_{\mathbb{A}}} \phi_s(1)$,
(T2) If $s_1, \ldots, s_k \in S_{\mathbb{A}}$ are powers of different primes then $\phi_{s_1 \cdots s_k}(x) | p_{\mathbb{A}}(x)$.

They show that if $\mathbb{A}$ satisfies these two conditions then $\mathbb{A}$ tiles the integers, and if $\mathbb{A}$ tiles and $|\mathbb{A}|$ has at most two prime factors then it satisfies *(T1)* and *(T2)*.

This is related to a more general conjecture of Fuglede [22]. In this case Fuglede's conjecture becomes the assertion that the set $\mathbf{A} = \{\bigcup_{a \in \mathbb{A}} [a, a+1) \subset \mathbb{R}\}$ tiles the real line if and only if the set $\mathbf{A}$ is a spectral set, meaning for some discrete set $\mathscr{T} \subset \mathbb{R}$ there is a set of exponentials $\{e^{2\pi i \lambda t} : t \in \mathscr{T}\}$ which forms an orthogonal basis of $L^2(\mathbf{A})$ ($\mathscr{T}$ is the spectrum).

7.1.3 Infinite $\mathbb{A}, \mathbb{B}$: No Structure Expected

When both sets $\mathbb{A}$ and $\mathbb{B}$ are infinite, no structure should be expected as the next two theorems show: the first due to Swenson [52] and the second due to Eigen and Hajian [11].

Theorem 7.1.3. *Let $\mathbb{A}$ and $\mathbb{B}$ be two finite sets of integers such that their sumset is direct. Then $\mathbb{A}, \mathbb{B}$ are extendable to infinite sets $\tilde{\mathbb{A}}, \tilde{\mathbb{B}}$ such that $\tilde{\mathbb{A}} \oplus \tilde{\mathbb{B}} = \mathbb{Z}$.*

Proof. Suppose $n \notin \mathbb{A} \oplus \mathbb{B}$. We will extend $\mathbb{A}$ and $\mathbb{B}$ so that the new sum is still direct and contains n. The result then follows by induction.

Without loss of generality we may assume $\mathbb{A} \cap \mathbb{B} = \{0\}$. Let us assume $\mathbb{A} \oplus \mathbb{B} \subset [-N, \ldots, N]$.

If $n > 0$ then let us put $2N + 1 + n$ in the set $\mathbb{A}$ and $-2N - 1$ in the set $\mathbb{B}$. Clearly, $2N + 1 + n - 2N - 1 = n$ so n is in the extended sum. Any number of the form $2N + 1 + n + b$ is greater than or equal to $N + 1 + n$ and so is disjoint from $\mathbb{A} \oplus \mathbb{B}$. Any number of the form $a + (-2N - 1) \leq -N - 1$ and so is also disjoint from both $\mathbb{A} \oplus \mathbb{B}$ and from $2N + 1 + n + b$. Thus the extensions are direct. If $n < 0$ a similar argument holds.   □

Theorem 7.1.4. *If a sequence of integers $\mathbb{A} = \{0 = a_0 < a_1 < a_2 < \cdots\}$ satisfies $\lim_{n \to \infty} (a_n - 2a_{n-1}) = \infty$, then there exists a sequence $\mathbb{B}$ such that $\mathbb{A} \oplus \mathbb{B} = \mathbb{Z}$; in other words, $\mathbb{A}$ tiles $\mathbb{Z}$.*

In the above the "2" is sharp in this sense. Define the sequence $a_i = \sum_{j=0}^i j$. Hence $a_{i+1} - a_i = i + 1 \to \infty$. However, $\mathbb{A} = \{a_i : i \geq 0\}$ does not have a complementing set in $\mathbb{Z}$ because the difference set $\mathbb{A} - \mathbb{A} = \mathbb{Z}$—as we will see, by Lemma 7.2.6, $\{0\}$ is the only sequence that has a direct sumset with $\mathbb{A} = \{a_i : i \geq 0\}$.

Proof. The following proof of Theorem 7.1.4 is due to J. Schmerl (Private communication, Department of Mathematics, University of Connecticut, Storrs). The result originally appeared in [11, 12] and it followed as a corollary to a result on infinite ergodic transformations.

Let $\{0 = z_0, z_1, z_2, \ldots, z_n, \ldots\}$ be an enumeration of elements in the set $\mathbb{Z}$. We define a sequence $\{\mathbb{B}_i\}$ of subsets of $\mathbb{Z}$ inductively as follows: $\mathbb{B}_0 = \{0\}$. Suppose the sets $\mathbb{B}_0 \subset \mathbb{B}_1 \subset \cdots \subset \mathbb{B}_i$ have been defined for $i \geq 0$. If $z_i \in A + \mathbb{B}_i$ then let $\mathbb{B}_{i+1} = \mathbb{B}_i$. For any set of integers $\mathbb{C}$ we define $\|\mathbb{C}\| = max\{|c| : c \in \mathbb{C}\}$. If $z_i \notin A + \mathbb{B}_i$ then let n_i be a sufficiently large positive integer such that for $j \geq n_i$

$$a_j - 2a_{j-1} > |z_i| + \|\mathbb{B}_i\|.$$

Then we let $\mathbb{B}_{i+1} = \mathbb{B}_i \cup \{z_i - a_{n_i}\}$. Next we let $\mathbb{B} = \cup_{n \geq 0} \mathbb{B}_n$. First we see that $\mathbb{Z} = A + \mathbb{B}$. Indeed if $z_i \in A + \mathbb{B}_i$ then $z_i \in A + \mathbb{B}$, while if $z_i \notin A + \mathbb{B}_i$ then by the definition of $\mathbb{B}_{i+1}$, $z_i - a_{n_i} \in \mathbb{B}_{i+1}$ so that $z_i = a_{n_i} + z_i - a_{n_i} \in A + \mathbb{B}_{i+1} \subset A + \mathbb{B}$.

Finally, we prove that the representation of the numbers in $\mathbb{Z}$ as $a + b$ with $a \in A$ and $b \in \mathbb{B}$ is unique by using induction on i, where the subscript i is in the construction of $\mathbb{B}_i$ above. Clearly, this is true for $i = 0$. Suppose that every number in the set $A + \mathbb{B}_i$ is uniquely represented as $a + b$ with $a \in A$ and $b \in \mathbb{B}_i$. If $\mathbb{B}_{i+1} = \mathbb{B}_i$ then the same is true for $A + \mathbb{B}_{i+1}$. Suppose that $z_i \notin A + \mathbb{B}_i$ and $\mathbb{B}_{i+1} = \mathbb{B}_i \cup \{z_i - a_{n_i}\}$. We need to show that $a_j + z_i - a_{n_i}$ cannot be represented in the form $a + b$ with $a \in A$ and $b \in \mathbb{B}_i$. Suppose that we have $a + b = a_j + z_i - a_{n_i}$ where $a \in \mathbb{Z}$ and $b \in \mathbb{B}_i$. We consider the following three cases separately:

Case (i): $j = n_i$.
   In this case $a + b = a_{n_i} + z_i - a_{n_i} = z_i$, but by assumption $z_i \notin A + \mathbb{B}_i$ and $b \in \mathbb{B}_i$; therefore $a \in A$ is impossible.
Case (ii): $j < n_i$.
   In this case we have $a_j < a_{n_i-1}$, and since $z_i - b \leq |z_i| + \|\mathbb{B}_i\|$ we conclude that

$$a = a_j + z_i - a_{n_i} - b \leq a_{n_i-1} - a_{n_i} + z_i - b$$
$$\leq |z_i| + \|\mathbb{B}_i\| - (a_{n_i} - 2a_{n_i-1}) - a_{n_i-1}$$
$$< -a_{n_i-1}$$
$$< 0.$$

Hence we have $a \notin A$ in this case also.
Case (iii): $j > n_i$.
   In this case we have $a_{j-1} \geq a_{n_i}$. Therefore

$$a = a_j + z_i - a_{n_i} - b \leq a_j + |z_i| + \|\mathbb{B}_i\| - a_{n_i}$$
$$< a_j,$$

and

$$a = a_j + z_i - a_{n_i} - b \geq a_j - a_{j-1} + z_i - b$$
$$= a_j - 2a_{j-1} + z_i - b + a_{j-1}$$
$$= a_j - 2a_{j-1} - (b - z_j) + a_{j-1}$$
$$\geq a_j - 2a_{j-1} - (|z_i| + \|\mathbb{B}_i\|) + a_{j-1}$$
$$> a_{j-1}.$$

Therefore we have $a_{j-1} < a < a_j$ which implies that $a \notin \mathbb{A}$ in this case also. Thus we conclude $\mathbb{Z} = \mathbb{A} \oplus \mathbb{B}$. $\square$

7.2 How Tilings Arise in Ergodic Theory

In this section we show how infinite complementing pairs of integers arise in infinite ergodic theory. Specifically, we will see that every *eww* sequence for a transformation is one member of a complementing pair in $\mathbb{Z}$. Additionally, almost every point in the *eww* set for that transformation gives rise to a complementing set for the *eww* sequence via its forward–backward hitting times. From this, we give a characterization of *eww* sequences. This is an extension of Kamae's [46] characterization of *ww* sequences.

Let T be an infinite ergodic transformation defined on $(X, \mathcal{B}, \mu)$. Let $A \in \mathcal{B}$ be an *eww* set and let $\mathbb{A}$ be its associated *eww* sequence. For each point $x \in A$ consider the following sequences of integers (the *hitting times* of x to A and the *return times* of x to A respectively):

$$\mathbb{H}(x) = \{n \in \mathbb{Z} : T^n x \in A\}. \tag{7.1}$$

$$\mathbb{R}(x) = \{n \in \mathbb{N} : T^n x \in A\}. \tag{7.2}$$

Theorem 7.2.1. *For a.a. $x \in A$, the sets $\mathbb{H}(x)$ and $\mathbb{A}$ are two infinite sets tiling $\mathbb{Z}$; i.e. $\mathbb{H}(x) \oplus \mathbb{A} = \mathbb{Z}$.*

Note that it is not necessary to have $x \in A$. This is just done to normalize the sequence so that $0 \in \mathbb{H}(x)$. Similarly, it is not necessary for $0 \in \mathbb{A}$, but by translating the set $\mathbb{A}$ we may assume $0 \in \mathbb{A}$.

Proof. We begin by defining what we mean by "generic points for A" for this proof. By throwing out from A, if necessary, a set of measure zero we assume that $T^w(A) \cap T^{w'}(A) = \emptyset$ for $w \neq w'$ and both $w, w' \in \mathbb{A}$. Let us denote by $\mathscr{G}$ the denumerable field generated by T and A, $\{T^n A : \text{for } -\infty < n < \infty\}$. We remove from $\mathscr{G}$ all the sets $G \in \mathscr{G}$ with $\mu(G) = 0$ and all their images $T^n G$.

Each remaining $G \in \mathcal{G}$ has positive measure and since T is ergodic almost all points in X enter G infinitely often. We remove from X the points which do not hit the remaining sets $G \in \mathcal{G}$ infinitely often. Altogether, we will have removed a countable number of sets of measure zero. We will refer to the remaining points as "generic."

Fix one of the remaining generic points $x \in A$. We now show that the directness of the sumset $\mathbb{A} + \mathbb{H}(x)$ will follow from the disjointness of the images of A under the ww sequence $\mathbb{A}$. Furthermore, that the direct sum $\mathbb{A} \oplus \mathbb{H}(x)$ is all of $\mathbb{Z}$ follows from the exhaustiveness of the set A under the sequence $\mathbb{A}$. The proofs of these two facts follow.

For $h, h' \in \mathbb{H}$ and $w, w' \in \mathbb{A}$ if $h + w = h' + w'$ then $T^{h+w}x = T^{h'+w'}x$, and it follows that $T^{h+w}x = T^w T^h x$ and $T^{h'+w'}x = T^{w'} T^{h'} x$ are the same point. By the definition of $\mathbb{H}(x)$ the two points $T^h x$ and $T^{h'} x$ are both in the set A. Hence $T^{h+w}x = T^{h'+w'}x$ is in a unique image of A in the partition of X given by $\{T^a A : a \in \mathbb{A}\}$. Thus $w = w'$, and it follows that $h = h'$, or the sumset is direct.

If $n \in \mathbb{Z}$ then $T^n x$ must be in one of the disjoint images $T^w A$. Applying T^{-w} to A we have $T^{-w} T^n x \in A$. By definition, $n - w \in \mathbb{H}(x)$. Therefore $n - w = h$, and $n = h + w \in \mathbb{H}(x) \oplus \mathbb{A}$. This shows $\mathbb{H}(x) \oplus \mathbb{A} = \mathbb{Z}$. $\qquad\square$

The converse is similar. As before, we suppose $A \in \mathcal{B}$ and T is an infinite ergodic transformation. For each $x \in A$ let $\mathbb{H}(x)$ be the hitting times to A.

Theorem 7.2.2. *If $\mathbb{H}(x) \oplus \mathbb{A} = \mathbb{Z}$ for almost all $x \in A$, then the set A is eww with the sequence $\mathbb{A}$.*

Proof. Suppose A is not ww with the sequence $\mathbb{A}$. Then we can find $w \neq w' \in \mathbb{A}$ such that $\mu(T^w A \cap T^{w'} A) > 0$. Assuming that $x \in A$, by the fact that T is ergodic there exists an integer n such that $T^n x = T^w a = T^{w'} a'$ for $a, a' \in A$. Then there exist $h, h' \in \mathbb{H}(x)$, where $T^h x = a$ and $T^{h'} x = a'$. Thus $T^n x = T^{w+h} x = T^{w'+h'} x$, which contradicts the directness assumption.

Suppose A is not exhaustive with $\mathbb{A}$, i.e. $\mu(X \setminus \cup_{w \in \mathbb{A}} T^w A) > 0$. T ergodic implies that there is an integer n with $T^n x \in X \setminus \cup_{w \in \mathbb{A}} T^w A$, and this n cannot be of the form $h + w$, contradicting the assumption that the sum $\mathbb{H}(x) + \mathbb{A}$ is all of $\mathbb{Z}$. $\qquad\square$

We note that the measurable set A will be ww with $\mathbb{A}$ if only one generic point x has hitting times $\mathbb{H}(x)$ with a direct sum; that is $\mathbb{H}(x) + \mathbb{A} = \mathbb{H}(x) \oplus \mathbb{A}$.

Corollary 7.2.3. *If there exists a generic point $x \in A$ such that the sumset of $\mathbb{A}$ and $\mathbb{H}(x)$ is direct then the set A is ww with the sequence $\mathbb{A}$.*

Proof. This follows from the definition of a generic point and the fact that T is ergodic. That is, if the set A is not ww with $\mathbb{A}$ then the orbit of the generic point would have to go through the intersection of the images of A which overlap. This contradicts the directness of $\mathbb{A}$ with the hitting times of the generic point. $\qquad\square$

From the above we see that the *eww* sequence $\mathbb{A}$ has a continuum of complements derived from the points $x \in A$ almost everywhere. Even though these sequences $\mathbb{H}(x)$ are different they do possess an additional common property. We recall the notation for the difference set.

Definition 7.2.4. The *difference set* of a set $\mathbb{A}$ is $\mathbf{D}(\mathbb{A}) = \{n - n' \, : \, n, n' \in \mathbb{A}\}$.

The proof of the next proposition is an easy consequence of T being ergodic.

Proposition 7.2.5. *Let $x \neq y \in A$ be generic points (as defined in Theorem 7.2.1). Then the difference sets of their hitting times are equal:*

$$\mathbf{D}(\mathbb{H}(x)) = \mathbf{D}(\mathbb{H}(y)).$$

It is also easy to see the following.

Lemma 7.2.6. *If $\mathbb{A}$ and $\mathbb{B}$ are direct, $(\mathbb{A} + \mathbb{B} = \mathbb{A} \oplus \mathbb{B})$ then $\mathbf{D}(\mathbb{A}) \cap \mathbf{D}(\mathbb{B}) = \{0\}$.*

In Sect. 7.3.7, we will show that there are transformations with *eww* sequences which have more than one *eww* set. It is possible for the hitting sequences of points in these other sets to be different, resulting in additional complementing sets for the *eww* sequence, while at the same time it is possible for the difference sets to be the same.

7.2.1   Constructing a Transformation from a Hitting Sequence

Now we illustrate how much more complicated and difficult the situation is when both members of a complementing pair are infinite. An obvious question raised by the previous section is whether every infinite complementing pair in $\mathbb{Z}$ arises as the hitting times of a generic point of an infinite ergodic transformation.

To deal with this question we make the following definitions which abstract the properties of the hitting and return times of a point to a set given in Eqs. (7.1) and (7.2).

Definition 7.2.7. An infinite monotone sequence of integers $\mathbb{H} = \{\cdots < h_{-1} < 0 = h_0 < h_1 < \cdots\}$ is called a *hitting sequence* if it has the following *double shift-repeat property*: for any $n \in \mathbb{Z}$ and any finite consecutive block $\{h_n, h_{n+1}, \ldots, h_{n+k}\}$ in $\mathbb{H}$, there exist $i, j > 0$ such that $\{h_n + h_i, h_{n+1} + h_i, \ldots, h_{n+k} + h_i\}$ and $\{h_n + h_{-j}, h_{n+1} + h_{-j}, \ldots, h_{n+k} + h_{-j}\}$ are again consecutive blocks of $\mathbb{H}$.

This is an extension of Kamae's returning sequence $\mathbb{R} = \{0 = r_0 < r_1 < \cdots\}$ which satisfies a shift-repeat property only in the positive direction. The notion of a hitting sequence in a certain sense characterizes *eww* sequences which arise in ergodic theory, and the following holds easily for generic points.

Remark 7.2.8. For a.a. points $x \in A$, $\mathbb{H}(x)$ the hitting times of x to A, satisfies the double shift-repeat property and so is a hitting sequence. We also note $\mathbb{R}(x)$ satisfies the shift-repeat property and thus is a returning sequence.

We would like to extend a result of Kamae [46] by characterizing *eww* sequences. But first we discuss a characterization of *ww* sequences (see [10]); its proof will illustrate the techniques involved for characterizing *eww* sequences.

Proposition 7.2.9. *A sequence* $\mathbb{A}$ *is ww for some ergodic transformation* T *if and only if there exists a hitting sequence* $\mathbb{H}$ *such that* $\mathbb{H} + \mathbb{A} = \mathbb{H} \oplus \mathbb{A}$.

Proof. We assume $\mathbb{H}$ and $\mathbb{A}$ are as above. We wish to construct a measure space X, a set A and an ergodic transformation T for which the set A is *ww* with $\mathbb{A}$ for T. Our construction will be one of cutting, stacking and inserting; i.e. at each step we will cut the previous pieces, stack them and insert additional pieces between the stacks. The transformation T will be the usual one which moves up the stack (see [21] for more details on cutting and stacking constructions). Note that no attempt is made to make T measure-preserving. If levels of different width are stacked we assume the transformation to be appropriately scaled between them.

We begin with $X = [0, 1)$ and let $A = [0, 1/2)$ and $E = [1/2, 1)$. The set A will be our base set; i.e. the *ww* set. It will be necessary to cut it up and stack it. The set E will be the extra piece (often referred to as the spacer set). It will be cut up and inserted between various pieces of A. All cut intervals will be left-closed and right-open. The construction will be arranged so that the point $0 \in [0, 1)$ will be generic and have $\mathbb{H}(0)$, the hitting times of 0 to A, satisfy $\mathbb{H}(0) = \mathbb{H}$. This will be enough to guarantee the disjointness of A under the $\mathbb{A}$ iterates. Let us write $\mathbb{H}$ as

$$\mathbb{H} = \{ \cdots < h_{-1} < 0 = h_0 < h_1 < \cdots \}.$$

We consider h_1 in $\mathbb{H}$. Thus the orbit of 0 should leave A and not return until the h_1 iterate of T. To achieve this we cut A in half. Now take the piece with 0 in it and cut that in half. We stack these last two pieces with 0 in the bottom piece. From E we remove half and cut this into $h_1 - 1$ pieces. These are inserted between the two stacked pieces of A (see Fig. 7.1).

Now by the shift-repeat property in the negative direction we can find $j > 0$ such that $h_{-j+1} - h_{-j} = h_1 - h_0$. We cut the stack in half and place the right side of the cut stack under the left (see Fig. 7.2). Between these two portions of the stack we must place more pieces of A. For each n, $-j + 1 < n < 0$, we put pieces of A at levels corresponding to the value of the hitting time h_n. Specifically we need $j - 2$ pieces corresponding to the hitting time values h_{-1} and h_{-j+2}. Between successive hitting values h_n and h_{n+1} we stack as many pieces of E as necessary to give the required hitting times (see Fig. 7.2).

We continue as follows. The block $\{h_{-j}, \ldots, h_1\}$ must be shift-repeated in the forward direction; i.e. we find $i > 0$ so that $\{h_{-j} + h_i, \ldots, h_1 + h_i\}$ is a consecutive block and such that $h_1 < h_{-j} + h_i$ (so the blocks do not overlap). This allows us to cut the stack in half and this time place the right half on top of the left. For each

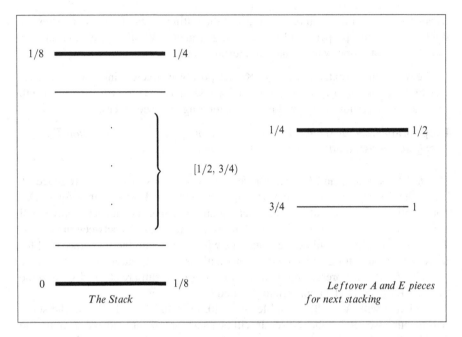

Fig. 7.1 For column on *left*: cut the left half of A in half and insert $h_1 - 1$ pieces from half of E

$h \in \mathbb{H}$, $h_1 < h < h_{-j} + h_i$, we must insert an additional piece of A between the two halves of the stack. Between each of these additional levels of A and the two halves of the stack we must insert additional levels of E. Specifically, between the h_k level of A and the h_{k+1} level of A we insert $h_{k+1} - h_k - 1$ pieces of E. The process continues inductively using up all of $\mathbb{H}$ and all of the sets A and E (see Fig. 7.3).

Clearly the hitting times of 0 in A is $\mathbb{H}$. In addition it is not hard to see that the set A is *ww* with the sequence $\mathbb{A}$. To see that the transformation so constructed is ergodic we observe that the induced transformation on A is ergodic, since the widths of the cuts are going to zero (see [21]).   □

The transformation constructed in Proposition 7.2.9 need not have the set A exhaustive with $\mathbb{A}$ even though it is *ww*. We modify the construction to make A *eww* with $\mathbb{A}$.

Theorem 7.2.10. *Let $\mathbb{A}$ be a sequence of integers such that there exists a hitting sequence $\mathbb{H}$ such that $\mathbb{H} \oplus \mathbb{A} = \mathbb{Z}$. Then there exists an infinite ergodic transformation and a set A of positive measure which is eww with $\mathbb{A}$.*

Proof. We begin as before with the cutting, stacking and inserting step. Assume we have constructed a stack S of A levels and E levels corresponding to the block $[h_{-j}, h_{-j}+1, \ldots, h_i]$. Since $\mathbb{H} \oplus \mathbb{A} = \mathbb{Z}$ we can find integers $n, m > 0$, $h_{-m} < h_{-j}$ and $h_i < h_n$, such that $\{h_{-m}, \ldots, h_n\} \oplus \mathbb{A} \supset [h_{-j}, h_{-j}+1, \ldots, h_i]$. For each $h \in \mathbb{H}$, $h_{k+1} - h_k - 1$, we place a piece of A below the stack S. Between these pieces of A

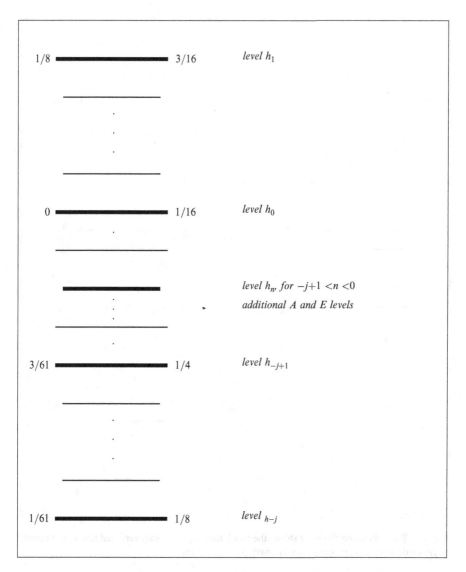

Fig. 7.2 Cut the first stack in half, and stack the right half under the left half. Between [3/16, 1/4) and [0,1/16) stack $j - 2$ pieces of A. Then between successive pieces of A stack pieces of E

corresponding to levels h_k and h_{k+1} we place $h_{k+1} - h_k - 1$ pieces of E. Similarly, we place additional pieces of A and E above S. This results in a new stack S'. At this stage we see that under $\mathbb{A}$, the images of A in S' cover the entire previous stack S. We continue the construction by alternating the cutting, stacking and inserting step with this exhausting step. $\qquad\qquad\square$

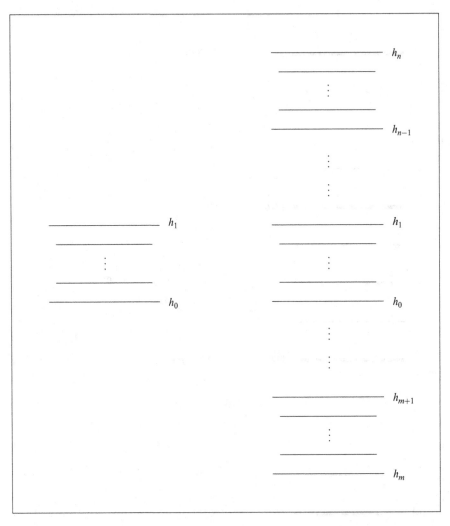

Fig. 7.3 The column on the *right* shows the block from h_0 to h_1 shift-repeated above and below. Between these blocks are appropriately many pieces of A and E

We now show how to modify the construction to yield a measure-preserving transformation. Obviously, this means at any stage all the new pieces we add should have the same size as the width of the current stack. However, we cannot simply start with a set A of infinite measure and cut it in half and stack it.

Theorem 7.2.11. *Let $\mathbb{A}$ be a sequence for which there exists a hitting sequence $\mathbb{H}$ such that $\mathbb{H} \oplus \mathbb{A} = \mathbb{Z}$. Then there exists an infinite ergodic transformation T and a set A of positive measure (possibly infinite measure) which is eww with $\mathbb{A}$.*

Proof. Let $I = [0, 1)$. The set I will be a part of the set A which we will construct. We assume there is an additional (disjoint from I) real line $\mathbb{R}$ from which we can take more pieces. Some of these will be used as inserts in the way we did with the set E (our spacer pieces), and some will be used to fill I out to the set A.

We again begin with h_1. This tells us to cut I in half and stack the right-hand side on top of the left. Between the two we insert $h_1 - 1$ additional intervals of the same size and mark these as pieces of E. Our stack at this step is of height $h_1 + 1$, and the set A consists of the top and bottom levels. These two h_0 and h_1 levels correspond to the two members of $\mathbb{H}$.

We perform all the previous cutting, stacking and inserting chores at the even steps of the induction—noting that all the new A pieces and E pieces are cut from the real line $\mathbb{R}$ and have the correct width. So even though the set A is enlarging (and its measure may well go to infinity) we guarantee that on the set I the induced transformation will be ergodic. At the odd steps we perform the exhausting step— adding extra A and E levels and insuring that the images of A under $\mathbb{A}$ cover all the A and E levels of the previous even step. Both steps maintain the disjointness of the set A and this proves the theorem. □

7.3 Examples of Complementing Pairs

In the construction of Swenson's Theorem 7.1.3 and in the construction of Theorem 7.1.4 neither set of the resulting complementing pair is expected to have any double shift-repeat property. But one can ask the following question. If $\mathbb{A}$ and $\mathbb{B}$ are both infinite sequences with $\mathbb{A} \oplus \mathbb{B} = \mathbb{Z}$, does there exist a hitting sequence $\mathbb{H}$ such that $\mathbb{A} \oplus \mathbb{H} = \mathbb{Z}$? In this section we present two examples that address this question.

The first example (due to J.D. Clemens) gives a sequence $\mathbb{A}$ which has complementing sets in $\mathbb{Z}$, but none of the complementing sets has the shift-repeat property. That is, $\mathbb{A}$ is not an *eww* sequence for any transformation.

The second example concerns an *eww* sequence $\mathbb{A}$ for a particular transformation. This sequence has a continuum of complementing sets indexed by the points of the *eww* set. Moreover, the transformation has more than one set which is *eww* under the sequence $\mathbb{A}$ and some of these sets give additional complementing sets via the hitting times of points to these other sets. The question then is: are there any complementing sets of $\mathbb{A}$ not given by the hitting times of a point to a set? The answer is yes, and the second example will show that despite being the *eww* sequence for a transformation, $\mathbb{A}$ still has a complementing set which does not satisfy the shift-repeat property.

Despite this negative result we will show in Sect. 7.4 that it is still possible in certain cases to characterize all the complementing sets of an *eww* sequence $\mathbb{A}$ using p-adic analysis.

7.3.1 A Complementing Set That Is Not a Hitting Sequence

In this section we present a result of Clemens [3]. Clemens's result applies to returning sequences as well as to hitting sequences, although we only state the result for hitting sequences (recall that a returning sequence is just the nonnegative return times for the orbit of some point to a set).

Theorem 7.3.1. *There exists a sequence* $\mathbb{A}$ *which has a complementing set in* $\mathbb{Z}$ *but has no complementing set which is a hitting sequence for any ergodic transformation.*

Therefore the sequence $\mathbb{A}$ cannot be *eww* or *ww* for any ergodic transformation. We begin with the following observation.

Lemma 7.3.2. *The difference set* $D(\mathbb{H})$ *of a hitting sequence always contains an arithmetic progression of length 3.*

Proof (Lemma 7.3.2). We assume the hitting sequence $\mathbb{H}$ contains 0, and let a be the next integer in the sequence. By the shift-repeat property there is a $b \in \mathbb{H}$ so that $b + a$ is also in $\mathbb{H}$. Consider the differences $b - a$, $b - 0$ and $(b + a) - 0$. They clearly form an arithmetic progression in $D(\mathbb{H})$.   □

Proof (Theorem 7.3.1). The sets $\mathbb{A}$, $\mathbb{B}$ such that $\mathbb{A} \oplus \mathbb{B} = \mathbb{Z}$ will be constructed inductively via an increasing sequence of finite sets $\mathbb{A}_n$ and $\mathbb{B}_n$ following Swenson's construction for Theorem 7.1.3.

There are essentially two parts to each inductive step.

- *Part 1:* Join a to $\mathbb{A}$ and b to $\mathbb{B}$, so that $a + b = k$ where k is an integer to be added to the sum $\mathbb{A} \oplus \mathbb{B}$ to guarantee exhaustiveness and without adding any arithmetic progressions of length 3 to the difference set of $\mathbb{B}$. This will be straightforward.
- *Part 2:* Join additional numbers to $\mathbb{A}$ for the purpose of making sure there can't be a hitting sequence $\mathbb{H}$ which is a complement of $\mathbb{A}$.

The ideas behind *Part 2* are:
- The sumset is direct, $\mathbb{A} + \mathbb{B} = \mathbb{A} \oplus \mathbb{B}$, if and only if $D(\mathbb{B}) \subset (\mathbb{Z} \setminus D(\mathbb{A})) \cup \{0\}$.
- If $(\mathbb{Z} \setminus D(\mathbb{A})) \cup \{0\}$ contains no arithmetic progression of length 3 then there cannot exist a hitting sequence $\mathbb{H}$ satisfying $\mathbb{A} + \mathbb{H} = \mathbb{A} \oplus \mathbb{H}$.

The following is the induction step for the construction.

We begin with $\mathbb{A}_0 = \mathbb{B}_0 = \{0\}$ and enumerate the integers in some order denoted by $0 = z_0, z_1, z_2, \dots$. At the n-th stage assume we have two finite sets $\mathbb{A}_n$ and $\mathbb{B}_n$ such that their sumset is direct, and such that the difference set $D(\mathbb{B}_n)$ of $\mathbb{B}_n$ contains no arithmetic progression of length 3.

If $z_{n+1} \in \mathbb{A}_n \oplus \mathbb{B}_n$ then we set $\mathbb{A}_{n+1} = \mathbb{A}_n$ and $\mathbb{B}_{n+1} = \mathbb{B}_n$.

Part 1

Suppose $z_{n+1} \notin \mathbb{A}_n \oplus \mathbb{B}_n$ and assume $\mathbb{A}_n \oplus \mathbb{B}_n \subset [-N, \dots, N]$. This means the largest difference in $D(\mathbb{B}_n)$ is at most $2N$. If $z_{n+1} > 0$ then put $\mathbb{B}_{n+1} =$

$\mathbb{B}_n \cup \{-3N-1\}$ and $\mathbb{A}' = \mathbb{A}_n \cup \{3N+1+z_{n+1}\}$. Hence $z_{n+1} \in \mathbb{A}' \oplus \mathbb{B}_{n+1}$ which is direct. It is also clear that all the new differences in $D(\mathbb{B}_{n+1})$ are greater (in absolute value) than $2N$ and so there are no arithmetic progressions of length 3.

Part 2

Now we want to extend $\mathbb{A}'$ to $\mathbb{A}_{n+1}$ with the goal of enlarging $D(\mathbb{A}_{n+1})$ to prevent $\mathbb{A}_{n+1}$ from having a complementing set in $\mathbb{Z}$ whose difference set contains a three-term arithmetic progression.

Assume $\mathbb{A}' \oplus \mathbb{B}_{n+1} \subset [-M, \ldots, M]$, $M \geq N$. If $D(\mathbb{A}') \cup D(\mathbb{B}_{n+1}) \supset [-M, \ldots, M]$ then we put $\mathbb{A}_{n+1} = \mathbb{A}'$.

Suppose instead $\{n_1, n_2, \ldots, n_k\} = [-M, \ldots, M] \setminus (D(\mathbb{A}') \cup D(\mathbb{B}_{n+1}))$.
Extend $\mathbb{A}'$ by joining the following:

$$\mathbb{A}_{n+1} = \mathbb{A}' \cup \{10M, 10M + n_1, \ 10^2 M, 10^2 M + n_2, \ \ldots, 10^k M, 10^k M + n_k\}.$$

Clearly, $D(\mathbb{A}_{n+1}) \cup D(\mathbb{B}_{n+1}) \supset [-M, \ldots, M]$ and $\mathbb{A}_{n+1}$ and $\mathbb{B}_{n+1}$ are direct. We let $\mathbb{A} = \cup_n \mathbb{A}_n$ and $\mathbb{B} = \cup_n \mathbb{B}_n$. Then we see that any other complementing set $\mathbb{C}$ of $\mathbb{A}$ must satisfy $D(\mathbb{C}) \subset D(\mathbb{B})$ and so cannot contain any arithmetic progressions of length 3. □

Remark 7.3.3. The previous example (as well as the original Swenson result) concerns extending two finite sets which are direct, to two infinite sets which tile $\mathbb{Z}$. On the other hand, if one has two infinite sets which are direct then it is easy to give examples where they cannot be extended to a pair of complementing sets of the integers. Instead of extending both, one can ask if it is possible to extend one of them. We now give a counterexample to this for *ww* sequences.

7.3.2 A ww Sequence Which Is Not eww for Any Transformation

Suppose $\mathbb{A}$ is a *ww* sequence for an ergodic transformation T. Then one can ask if it is *eww* for T or even for a different transformation S. In this section we give an example of an infinite sequence which is *ww* for some ergodic transformation but cannot be *eww* for any transformation [3].

Theorem 7.3.4. *There exists a sequence $\mathbb{A}$ which is ww for some ergodic transformation but is not eww for any ergodic transformation.*

The result which guarantees that $\mathbb{A}$ is not *eww* for some transformation is the following:

Lemma 7.3.5. *Suppose $\mathbb{A}$ satisfies:*

1. $0 \in \mathbb{A}$.
2. $-1 \notin \mathbb{A}$.
3. For all $a \in \mathbb{A}$, we have $a + 1 \in D(\mathbb{A}) = \mathbb{A} - \mathbb{A}$.

Then there does not exist $\mathbb{B}$ such that $\mathbb{A} \oplus \mathbb{B} = \mathbb{Z}$.

This of course implies there is no hitting sequence which is a complementing set of $\mathbb{A}$, and therefore $\mathbb{A}$ cannot be *eww* for any transformation.

Proof (Lemma 7.3.5). Without loss of generality we assume that $\mathbb{A} \cap \mathbb{B} = \{0\}$. If $\mathbb{A} \oplus \mathbb{B} = \mathbb{Z}$ then there exist $a \in \mathbb{A}$ and $b \in \mathbb{B}$ such that $a + b = -1$. So $-b = a + 1$. By assumption 2, $b \neq 0$. Assumption 3 implies $a + 1$ belongs to $\boldsymbol{D}(\mathbb{A})$. Hence $-b - 0 = a + 1 \in \boldsymbol{D}(\mathbb{A}) \cap \boldsymbol{D}(\mathbb{B})$. This contradicts Lemma 7.2.6 which says the intersection should just be $\{0\}$.   □

Now we prove Clemens's Theorem 7.3.4.

Proof (Theorem 7.3.4). We prove this theorem by inductively constructing $\mathbb{A}$. At the same time we construct a hitting sequence $\mathbb{B}$ so that $\mathbb{A} \oplus \mathbb{B} \neq \mathbb{Z}$. In order to insure $-1 \notin \mathbb{A}$ we will show $\mathbb{A} \oplus \mathbb{B} = \mathbb{Z} \backslash \{-1\}$.

We begin the construction by induction starting with $\mathbb{A}_0 = \{0\} = \mathbb{B}_0$.

Let $\mathbb{K} = \{k_i\}_{i \geq 1}$ be an enumeration of all the integers except -1. Suppose at step n we have two finite sets $\mathbb{A}_n, \mathbb{B}_n$ such that:

1. $\mathbb{A}_n + \mathbb{B}_n = \mathbb{A}_n \oplus \mathbb{B}_n$ (the sumset is direct).
2. $-1 \notin \mathbb{A}_n \oplus \mathbb{B}_n$.
3. $k_i \in \mathbb{A}_n \oplus \mathbb{B}_n$ for $1 \leq i \leq n$.
4. $a + 1 \in \boldsymbol{D}(\mathbb{A}_n)$ for $-n \leq a \leq n, a \in \mathbb{A}_n$.
5. The block $\{b \in \mathbb{B}_n : -n \leq b \leq n\}$ shift-repeats in $\mathbb{B}_n$ in both directions.

There are three parts to each induction step.

Part 1 Join integers to $\mathbb{A}_n$ to insure k_{n+1} will be in the sum $\mathbb{A}_{n+1} \oplus \mathbb{B}_n$.
Part 2 Join integers to $\mathbb{A}_n$ to insure that $a + 1$ is in $\boldsymbol{D}(\mathbb{A}_{n+1})$.
Part 3 Join integers to $\mathbb{B}_n$ to insure the shift-repeat property holds for $\mathbb{B}_{n+1}$.

Part 1. In this part we will extend $\mathbb{A}_n$ to $\mathbb{A}'_n$ and $\mathbb{B}_n$ to $\mathbb{B}'_n$ retaining the directness of the new sumset.

For $N > 1$ assume $-N < a + b < N$ for $a \in \mathbb{A}_n$ and $b \in \mathbb{B}_n$. If $k_{n+1} \in \mathbb{A}_n \oplus \mathbb{B}_n$ then we don't need to join any integers to $\mathbb{A}_n$. If $k_{n+1} \notin \mathbb{A}_n \oplus \mathbb{B}_n$, $k_{n+1} > 0$, then adjoin $2N + 1 + k_{n+1}$ to $\mathbb{A}_n$ and $-2N - 1$ to $\mathbb{B}_n$. Hence k_{n+1} has been added to the new sumset. Observe that -1 has not been added to the new sum. The new sums are $(2N + 1 + k_{n+1} + b) > (2N + 1 + k_{n+1}) - N = N + 1 + k_{n+1} > 1$ and $a + (-2N - 1) < N + (-2N - 1) = -N - 1 < -1$. It is clear that $\mathbb{A}'_n \oplus \mathbb{B}'_n$ is a direct sum. A similar argument holds if $k_{n+1} < 0$.

Part 2. In this part we extend $\mathbb{A}'_n$ to $\mathbb{A}_{n+1}$.

Consider $\{a \in \mathbb{A}'_n : -(n + 1) \leq a \leq (n + 1)\}$. At most two of these do not satisfy $a + 1 \in \boldsymbol{D}(\mathbb{A}'_n)$. They would be $n + 1$ and $-n - 1$ respectively.

Assume $-M < a + b < M$ for $a \in \mathbb{A}'_n, b \in \mathbb{B}'_n$ with $M > 1$.

If we have $n + 1 \in \mathbb{A}'_n$ then we will join $(2M + 1)$ and $(2M + 1) + (n + 1) + 1$ to $\mathbb{A}'_n$. If we have $-n - 1 \in \mathbb{A}'_n$ then we join $(-2M - 1)$ and $(-2M - 1) + (-n - 1) - 1$ to $\mathbb{A}'_n$.

This obviously puts the $a + 1$ terms into $\boldsymbol{D}(\mathbb{A})$. It is also easy to see, as in previous arguments, that $\mathbb{A}_{n+1} \oplus \mathbb{B}'_n$. Finally, we observe as before, that $-1 \notin \mathbb{A}_{n+1} \oplus \mathbb{B}'_n$.

Part 3. In this part we extend $\mathbb{B}'_n$ to $\mathbb{B}_{n+1}$.

For $N > 1$ assume $-N < a + b < N$ for $a \in \mathbb{A}_{n+1}$, $b \in \mathbb{B}'_n$. For each $b \in \mathbb{B}'_n$ satisfying $-(n+1) \leq b \leq (n+1)$ join to $\mathbb{B}'_n$ both $(b+3N+1)$ and $(b-3N-1)$. This gives $\mathbb{B}_{n+1}$.

Writing $\mathbb{A} = \cup_n \mathbb{A}_n$ and $\mathbb{B} = \cup_n \mathbb{B}_n$ completes the proof.   □

7.3.3 An eww Sequence with a Complementing Set That Does Not Come from a Point

We know by example that there are sequences which tile the integers which are not *eww* for any ergodic transformation. Now we ask the following: if a sequence is *eww* for some transformation are all of its complementing sets the hitting sequence of some point? In this section we show by an example that the answer is no (see [17]).

We recall the *eww* sequence for the First Basic Example in Chap. 4. It is the set of sums of finite subsets of odd powers of 2, and let us denote it by $\mathbb{A}$;

$$\mathbb{A} = SFS\{2^{2i+1} : i = 0, 1, 2, \ldots\}.$$

We note that by convention, 0 being the sum over the empty subset, belongs to $SFS\{2^{2i+1} : i \geq 0\}$. Observe the contrast between Theorem 7.3.6 and Theorem 7.1.4.

Theorem 7.3.6. *Let $0 = k_1 < k_2 < \cdots$ be an increasing sequence of nonnegative integers. Then there exists a complementing set $\mathbb{C} = \{c_n : n = 0, 1, 2 \cdots\}$ (in $\mathbb{Z}$) of $\mathbb{A} = SFS\{2^{2i+1} : i = 0, 1, 2, \ldots\}$ with the property that $|c_n| - |c_{n-1}| > k_n$ for all n. Such a complementing set does not have the shift-repeat property and so cannot be the hitting times of any point to a set.*

It is tempting to conclude from this theorem that the complementing sets of $\mathbb{A}$ in $\mathbb{Z}$ have no common structure at all. This is not the case. In Sect. 7.4 we will fully describe the structure for all complementing sets of $\mathbb{A}$. Further we will be able to determine if a finite set which is direct with this $\mathbb{A}$ extends to a complementing set or not.

In order to prove Theorem 7.3.6 we reproduce the construction of the transformation T of the First Basic Example as a cutting, stacking and inserting construction. Along the way we will demonstrate the existence of many *eww* sets for T with the sequence $\mathbb{A}$.

We begin with the set $W = [0, 1)$ and consider W as the column C_0 of height $h_0 = 1$ and width $w_0 = 1$. We cut the column in half, place the right half over the left half and double the height by adding 2^1 spacers on top. Then at stage $n = 1$ we have a column C_1 of height $h_1 = 2^2$ and width $w_1 = 2^{-1}$. We refer to C_1 as the column of rank 1.

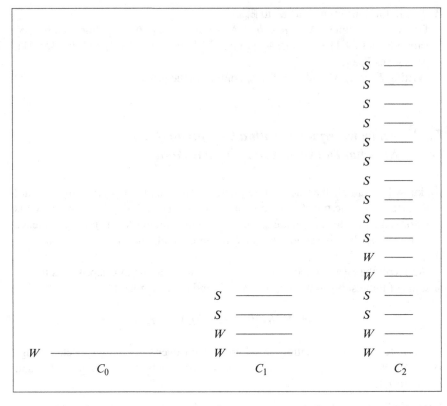

Fig. 7.4 Column construction of the first two columns: cut column in half, restack and double the space with spacers

Suppose at the $k - 1$ stage we have a column C_{k-1} of height $h_{k-1} = 2^{2k-2}$ consisting of intervals of width $w_{k-1} = 1/2^k$. We cut the column C_{k-1} in half, place the right side above the left side and add $2h_{k-1} = 2^{2k-1}$ spacers above. We refer to C_k as the column of rank k and note that it has height $h_k = 2^{2k}$ and width $1/2^{k+1}$.

The space $X = \cup C_k$ and the transformation T is the common map which goes up the columns linearly.

Proposition 7.3.7. *The transformation T (described by cutting and stacking) is isomorphic to the First Basic Example given earlier. The set $W = [0, 1)$ is an eww set under the sequence $\mathbb{A} = SFS\{2^{2i+1} : i \geq 0\}$.*

The results we use to prove Theorem 7.3.6 follow.

Proposition 7.3.8. *For positive integers k let L be a level of column C_k. Then*

$$T^{2^{2k-1}}\left(\bigcup_{a \in \mathbb{A}} T^a L\right) = \bigcup_{a \in \mathbb{A}} T^a L.$$

Proof. The proof is by induction on k, the rank of the column (Fig. 7.4).

For $k = 1$ the column C_1 has four levels denoted by L_1^i, $i = 0, \ldots, 3$. We observe that the first two levels of the column are actually two disjoint images of the *eww* set $W = L_0^0 = L_1^0 \cup L_1^1 = L_1^0 \cup TL_1^0$. Hence we have

$$X = \left(\bigcup_{a \in \mathbb{A}} T^a L_1^0 \right) \cup T \left(\bigcup_{a \in \mathbb{A}} T^a L_1^0 \right) \, (disj). \tag{7.3}$$

Applying T to both sides of (7.3) we get

$$X = T \left(\bigcup_{a \in \mathbb{A}} T^a L_1^0 \right) \cup T^2 \left(\bigcup_{a \in \mathbb{A}} T^a L_1^0 \right) \, (disj),$$

from which we conclude

$$T^2 \left(\bigcup_{a \in \mathbb{A}} T^a L_1^0 \right) = \left(\bigcup_{a \in \mathbb{A}} T^a L_1^0 \right) \, (disj).$$

Since all levels of column C_1 are images of the first level L_1^0 the result follows for these levels by repeatedly applying T.

Before proceeding with the proof we pause to point out that L_1^0 is part of the *eww* set W, while $T^2 L_1^0$ is one of the spacers added in the construction. This is proved as Corollary 7.3.9 following the proof of this proposition.

Next we return to the proof of Proposition 7.3.8 and assume that the proposition is true for k. That is, $T^{2^{2k-1}} \left(\bigcup_{a \in \mathbb{A}} T^a L \right) = \bigcup_{a \in \mathbb{A}} T^a L$ for all levels L in the column C_k.

By construction the first level L_k^0 of column C_k is a disjoint union of two levels of column C_{k+1}, the second of which is a T-image of the first. That is,

$$L_k^0 = L_{k+1}^0 \cup L_{k+1}^{2^{2k}} = L_{k+1}^0 \cup T^{2^{2k}} L_{k+1}^0;$$

from this we obtain

$$\bigcup_{a \in \mathbb{A}} T^a L_k^0 = \left(\bigcup_{a \in \mathbb{A}} T^a L_{k+1}^0 \right) \cup T^{2^{2k}} \left(\bigcup_{a \in \mathbb{A}} T^a L_{k+1}^0 \right) \, (disj).$$

Applying $T^{2^{2k}}$ to both sides and using the induction hypothesis we get

$$\bigcup_{a \in \mathbb{A}} T^a L_k^0 = T^{2^{2k}} \left(\bigcup_{a \in \mathbb{A}} T^a L_{k+1}^0 \right) \cup T^{2^{2k+1}} \left(\bigcup_{a \in \mathbb{A}} T^a L_{k+1}^0 \right) \, (disj).$$

From which we conclude $T^{2^{2k+1}} \left(\bigcup_{a \in \mathbb{A}} T^a L_{k+1}^0 \right) = \bigcup_{a \in \mathbb{A}} T^a L_{k+1}^0$. As each level in C_{k+1} is an image of the first level the proposition follows. $\quad\square$

In the previous proof we noted the following result.

Corollary 7.3.9. *The set* $U = L_1^0 \cup T^2 L_1^1$ *is another eww set for* T *with the sequence* $\mathbb{A}$.

Proof. The *eww* set $W = L_1^0 \cup L_1^1 (disj)$ satisfies

$$X = \bigcup_{a \in \mathbb{A}} T^a W$$

$$= \Big(\bigcup_{a \in \mathbb{A}} T^a L_1^0 \Big) \cup \Big(\bigcup_{a \in \mathbb{A}} T^a L_1^1 \Big) \, (disj)$$

$$= \Big(\bigcup_{a \in \mathbb{A}} T^a L_1^0 \Big) \cup \Big(\bigcup_{a \in \mathbb{A}} T^a T^2 L_1^1 \Big) \, (disj).$$

$\square$

We need the following lemmas before giving the proof of Theorem 7.3.6.

Lemma 7.3.10. *Let* U *be an eww set with the sequence* $\mathbb{A}$ *for* T, *and* V *a subset of* U. *Let* U *and* V *both be the finite union of levels of some column, and let* $k > 0$ *be a positive integer. Then there exists an integer* $j > k$ *such that the set* $E = (U \backslash V) \cup T^j V$ *is an eww set for the sequence* $\mathbb{A}$, *and the set* $T^j V$ *is at least* k *iterates away from* $U \backslash V$.

Proof. We note that according to Proposition 7.3.7 above and Theorem 5.1.2 the set U is necessarily of measure 1. Since U is a finite union of interval levels, by splitting these intervals further we represent both sets U and V as a finite union of disjoint levels of the column C_i of rank i for some $i > 0$. Since we can always use a column of higher rank we choose i such that the height $h_i = 2^{2i} > k$. We observe that at the next step $i + 1$ of the construction the sets U and V will be wholly contained in the bottom half of column C_{i+1}, i.e. 2^{2i+1} levels below the top. Continuing, we see that in column C_{i+2} the set U is 2^{2i+1} levels below the middle of the column. We put $j = 2^{2i+3}$; then the set $E = T^j V$ will be the finite union of complete levels (intervals), and it will be contained totally inside the top half of column C_{i+3} and 2^{2i+1} levels from the top of C_{i+3}. This says that the set $T^j V$ will be at least $2^{2i+1} > k$ iterates away from U. It is clear that if we let $E = (U \backslash V) \cup T^j V$ then E is *eww* for the sequence $\mathbb{A}$. $\square$

As noted earlier, the hitting times of almost all points in the *eww* set give a complementing set for the *eww* sequence. We now illustrate a "duality." Instead of changing the point we will change the set. That is, we fix a point and look at its hitting times to different sets.

Because we are fixing a point and varying the *set* (and consequently the *eww* sequence) we need to clarify what we mean by "generic" points. Thus in this section we define the *generic points* to be generated by the images of all the levels of the columns of W, i.e. in the field generated by $\{T^n W\}$. That is, we remove points which do not hit all the levels infinitely often. This is the same as saying the points are "generic" for the *eww* set W. As we modify W to a new set W', this new set

W' will consist of different levels of the original columns for W. Hence the points which are generic for W will also be generic for W'.

We fix a generic point x_0. Since the point is now fixed we denote the hitting times by $\mathbb{H}_W$ to emphasize the set W the point is hitting.

The next two results are immediate corollaries of Lemma 7.3.10 and we do not prove them.

Lemma 7.3.11. *Let x_0 be the designated generic point in W. Then for any integer $k > 0$ there is another eww set W' under the sequence $\mathbb{A}$ such that:*

1. *$x_o \in W'$,*
2. *W' is a finite union of disjoint levels of some column for the original W,*
3. *for all $h \in \mathbb{H}_{W'}$ if $h \neq 0$ then we have $|h| > k$.*

Proof. This follows directly from Lemma 7.3.10. □

Lemma 7.3.12. *Let x_0 be the designated generic point in W. Let W' be an eww set for $\mathbb{A}$ which consists of a finite union of levels of a column of the original set W. Let $x_0 \in W'$. Then for any finite set $\{0, n_1, \ldots, n_s\} \subset \mathbb{H}_{W'}$ and any $k > 0$, there exists another eww set W'' for $\mathbb{A}$ such that:*

1. *$x_0 \in W''$,*
2. *W'' is a finite union of levels of some column for the original W,*
3. *$\{0, n_1, \ldots, n_s\} \subset \mathbb{H}_{W''}$,*
4. *$|h| > k$ for all $h \in \mathbb{H}_{W''} \setminus \{0, n_1, \ldots, n_s\}$.*

Now we prove Theorem 7.3.6.

Proof (Theorem 7.3.6). We inductively choose the sequence $\mathbb{C} = \{c_n \in \mathbb{Z} : n = 0, 1, 2, \ldots\}$. Let $c_0 = 0$ and enumerate the integers as $\mathbb{Z} = \{z_n : n \geq 0\}$ with $z_0 = 0$.

Using Lemma 7.3.11 we choose W_1 such that for all $h \neq 0 \in \mathbb{H}_{W_1}$ we have $|h| > k_1$. Since x_0 is generic in W_1 there must exist a unique $a \in \mathbb{A}$ and $c_1 \in \mathbb{H}_{W_1}$ so that $z_1 = a + c_1$.

Next we use Lemma 7.3.12 and choose W_2 so that for all $h \in \mathbb{H}_{W_2}$ with $h \notin \{0, c_1\}$ we have $|h| > k_2 + |c_1|$. Let z be the next element in order in the enumeration $\{z_n\}$ which is not in $\{0, c_1\} \oplus \mathbb{A}$. Again x_0 is generic in W_2, and therefore there is $a \in \mathbb{A}$ and $c_2 \in \mathbb{H}_{W_2}$ so that $z = a + c_2$.

We proceed by induction. Having chosen the integers $\{0, c_1, \ldots, c_n\}$ we choose W_{n+1} so that $|h| > k_{n+1} + \max\{0, |c_1|, \ldots, |c_n|\}$ for all $h \in \mathbb{H}_{W_{n+1}}$ and $h \notin \{0, c_1, \ldots, c_n\}$. We let z be the next integer in the enumeration $\{z_n\}$ which is not in $\{0, c_1, \ldots, c_n\} \oplus \mathbb{A}$. Then we find a and $c_{n+1} \in \mathbb{H}_{W_{n+1}}$ with $z = a + c_{n+1}$. Continuing this way, the set $C = \{c_n : n = 0, 1, \ldots\}$ has the stated properties. □

As a final observation we mention the following corollary to the construction.

Corollary 7.3.13. *For the eww sequence $\mathbb{A}$ and the First Basic Example there is a continuum of eww sets.*

7.4　Extending a Finite Set to a Complementing Set

We continue our examination of the complementing sets (in $\mathbb{Z}$) of $\mathbb{A}$, the *eww* sequence for the First Basic Example T from Chap. 4. The results of this section will apply to $\mathbb{A}$ as a member of a larger class of sequences; see also [16].

In Sect. 7.1.2 we saw that when one of the complementing sets in a pair is finite there are structural results which deal with the question of when the finite set can tile the integers. When neither set is finite we saw in Sect. 7.1.3 that there is no general structure expected. Our investigation in this section focuses on a different type of question.

We begin with an infinite set of integers $\mathbb{C}$ which is known to tile the integers; that is, it has at least one complementing set. We consider two questions:

1. Can all the complementing sets of $\mathbb{C}$ be characterized?
2. Given a finite set of integers $\mathbb{F}$ which is direct with $\mathbb{C}$, can $\mathbb{F}$ be extended to a complementing set of $\mathbb{C}$ in $\mathbb{Z}$?

From the results in Sect. 7.1.3 the answer is no (no structure is expected). However, as an application we will ask these questions when $\mathbb{C}$ is the *eww* sequence for some infinite ergodic transformation. This now becomes a tractable problem.

We show that it is sometimes possible to answer these questions in an affirmative manner. Specifically, we work with $\mathbb{A} = SFS\{2^{2i+1} : i = 0, 1, 2, \ldots\}$, the sums of finite subsets of odd powers of 2, the *eww* sequence for the First Basic Example T.

To illustrate, it is easy to see that each of the sets $\{0, 18\}$, $\{0, 28\}$, $\{0, 52\}$ and $\{0, 28, 52\}$ have a direct sum with the set $\mathbb{A}$. We ask whether any of these four sets can be extended to a set $\mathbb{C}$ such that $\mathbb{A} \oplus \mathbb{C} = \mathbb{Z}$. It turns out that each of the sets $\{0, 28\}$ and $\{0, 52\}$ is extendable, while neither of the sets $\{0, 18\}$ or $\{0, 28, 52\}$ is. In the theorem that follows we give a necessary and sufficient condition for a finite set $\mathbb{F}$ of integers to be extendable to a set $\mathbb{C}$ which is a complementing set of $\mathbb{A}$ in $\mathbb{Z}$.

7.4.1　Definitions and Notations

Once again we consider the *eww* sequence $\mathbb{A}$ for the First Basic Example T; namely the set of all sums of finite subsets of the odd powers of 2,

$$\mathbb{A} = SFS \{2^{2i+1} : i = 0, 1, 2, \ldots\}$$

We also denote the corresponding set of integers for the even powers of 2 by

$$\mathbb{B} = SFS \{2^{2i} : i = 0, 1, 2, \ldots\}.$$

It is immediate that $\mathbb{A} \oplus \mathbb{B} = \mathbb{N}$. Furthermore, it is easy to see that $\mathbb{A} \oplus (-\mathbb{B}) = \mathbb{Z}$.

We generalize the preceding remarks. Let $\{\varepsilon_i : i = 0, 1, 2, \ldots\}$ be a sequence where $\varepsilon_i = 1$ or -1 for each i and define $\mathbb{F}(\{\varepsilon_i\}) = SFS\{\varepsilon_i 2^i : i = 0, 1, 2, \ldots\}$. We remind the reader that by convention 0 is always included in the SFS of a set and so is in $\mathbb{F}(\{\varepsilon_i\})$ for every sequence $\{\varepsilon_i\}$.

We claim that if $\varepsilon_i = +1$ for infinitely many i and also $\varepsilon_i = -1$ for infinitely many i then $\mathbb{F}(\{\varepsilon_i\})$ is equal to $\mathbb{Z}$. To see this we analyze $\mathbb{F}(\{\varepsilon_i\})$, by first defining $\mathbb{F}_n = SFS\{\varepsilon_i 2^i : i = 0, 1, 2, \ldots, n\}$. It is clear that $\mathbb{F}_0 \subset \mathbb{F}_1 \subset \mathbb{F}_2 \subset \cdots$ with $\bigcup \mathbb{F}_n = \mathbb{F}(\{\varepsilon_i\})$.

Lemma 7.4.1. *The set $\mathbb{F}_n$ consists of 2^{n+1} consecutive integers.*

Proof. Every element in $\mathbb{F}_n$ has a unique representation as a finite sum of the elements in the set $\{\varepsilon_i 2^i : i = 0, 1, 2, \cdots, n\}$. Hence the cardinality of the set $\mathbb{F}_n$ is 2^{n+1}. This coincides with the number

$$\max\{b \in \mathbb{F}_n\} - \min\{b \in \mathbb{F}_n\} + 1 = \sum_{i \in I_+} 2^i - \sum_{i \in I_-}(-2^i) + 1 = \sum_{i=0}^{n} 2^i + 1,$$

where $I_+ = \{i : 0 \le i \le n, \varepsilon_i = +1\}$ and $I_- = \{i : 0 \le i \le n, \varepsilon_i = -1\}$. $\square$

In what follows we choose a sequence $\{\varepsilon_i = \pm 1, i \ge 0\}$ such that $\varepsilon_i = +1$ for infinitely many i and $\varepsilon_i = -1$ for infinitely many i also. We fix this choice of the sequence $\{\varepsilon_i\}$ and denote from now on by $\mathbb{A}$ the $SFS\{\varepsilon_i 2^i : i \ge 0, i \text{ odd}\}$ and by $\mathbb{B}$ the $SFS\{\varepsilon_i 2^i : i \ge 0, i \text{ even}\}$. It follows that $\mathbb{A} \oplus \mathbb{B} = \mathbb{Z}$. We note that the case with $\varepsilon_{2i+1} = +1$ for all i corresponds to the *eww* sequence for the First Basic Example.

For any set of integers $\mathbb{E}$, we denote by $\mathscr{C}(\mathbb{E})$ the collection of complementing sets of $\mathbb{E}$ containing 0:

$$\mathscr{C}(\mathbb{E}) = \{\mathbb{C} \subset \mathbb{Z} : 0 \in \mathbb{C} \text{ and } \mathbb{E} \oplus \mathbb{C} = \mathbb{Z}\}.$$

For an integer $n \in \mathbb{Z}$ we denote by $ord_2(n)$ the highest power of 2 that divides n. Because of the special role that 0 has in this context we shall allow ∞ as the value for $ord_2(0)$ and consider 0 to be both of odd and even ord_2.

7.4.2   Extension Theorem

The following theorem establishes a necessary and sufficient condition for a finite set $\mathbb{F}$ to be extendable to a complementing set of $\mathbb{A}$.

Theorem 7.4.2 (Finite Extension). *Let $\mathbb{A} = SFS\{\varepsilon_i 2^i : i \ge 0, i \text{ odd}\}$, and let $\mathbb{F} \subset \mathbb{Z}$ be a finite set with $0 \in \mathbb{F}$. Then there is a complementing set $\mathbb{C}$ of $\mathbb{A}$ (that is a set $\mathbb{C} \in \mathscr{C}(\mathbb{A})$) with $\mathbb{F} \subset \mathbb{C}$ if and only if $ord_2(n)$ is even for any $n \in \mathbb{F} - \mathbb{F}$.*

We will prove the above using purely algebraic arguments. However, the origins of this result come from ergodic theory. Examine Fig. 6.1 for the First Basic Example. The bottom row contains a series of rectangles each containing a number. These numbers are forward return times of the left-hand endpoint $0 \in [0, 1)$ to the *eww* base set $W = [0, 1)$. First we observe that $ord_2(n)$ is even for every one of these return times. Then we observe that the difference $n - n'$ of any pair of return times also has $ord_2(n - n')$ even.

In Fig. 6.2 a second *eww* set is shown. The numbers in the rectangles again show the return times of the point 0 to this set. Examining these return times we still see that they all have even ord_2 and all the pairwise differences have even ord_2.

For any set $\mathbb{F}$ satisfying the conditions of the Theorem 7.4.2 there exists a hitting sequence $\mathbb{H}$ so that $\mathbb{F} \subset \mathbb{H}$; $\mathbb{H} = \mathbb{H}(0)$, the hitting times of 0 to W (the interval $[0, 1)$).

For the proof of the Finite Extension Theorem we need some preliminary lemmas; we also introduce some new concepts and notation to make the proof of the theorem more manageable.

Lemma 7.4.3. *Let* $\mathbb{B} = SFS \{\varepsilon_i 2^i : i \geq 0, i \text{ even}\}$. *For an even integer* $k \geq 0$, *let* $\mathbb{B}_k = SFS \{\varepsilon_i 2^i : i \geq 0, i \text{ even}, i \neq k\}$. *For* $\mathbb{D} \in \mathscr{C}(\mathbb{B})$ *let* $\hat{\mathbb{B}}_k = \mathbb{B}_k \oplus \mathbb{D}$. *Then* $\hat{\mathbb{B}}_k = \hat{\mathbb{B}}_k + n$ *for any integer* n *with* $ord_2(n) > k$.

Proof. We have

$$\mathbb{B} = (\mathbb{B}_k + \varepsilon_k 2^k) \cup \mathbb{B}_k \quad (disj);$$

therefore

$$\mathbb{Z} = (\hat{\mathbb{B}}_k + \varepsilon_k 2^k) \cup \hat{\mathbb{B}}_k \quad (disj). \tag{7.4}$$

Adding $\varepsilon_k 2^k$ to both sides of (7.4) we get

$$\mathbb{Z} = (\hat{\mathbb{B}}_k + \varepsilon_k 2^{k+1}) \cup (\hat{\mathbb{B}}_k + \varepsilon_k 2^k) \quad (disj). \tag{7.5}$$

Combining (7.4) and (7.5) we conclude

$$\hat{\mathbb{B}}_k = \hat{\mathbb{B}}_k + \varepsilon_k 2^{k+1}. \tag{7.6}$$

If $n \in \mathbb{Z}$ and $ord_2(n) > k$ then $n = p2^{k+1}$ for some integer p. From (7.6) then we conclude $\hat{\mathbb{B}}_k = \hat{\mathbb{B}}_k + n$. $\square$

We need a generalization of Lemma 7.4.3 to a finite set $0 \leq k_1 < k_2 < \cdots < k_p$ of even integers. The proof of the next lemma is accomplished by a straightforward induction argument along the lines of the proof of Lemma 7.4.3; we omit the proof.

Lemma 7.4.4. *Let* $0 \leq k_1 < k_2 < \cdots < k_p$ *be a finite set of even integers and let* $\mathbb{B}_{k_1 k_2 \cdots k_p} = SFS \{\varepsilon_i 2^i : i \geq 0, i \text{ even}, i \neq k_1, k_2, \ldots, k_p\}$. *For* $\mathbb{D} \in \mathscr{C}(\mathbb{B})$ *let*

$\hat{\mathbb{B}}_{k_1k_2\cdots k_p} = \mathbb{B}_{k_1k_2\cdots k_p} \oplus \mathbb{D}$. *Then* $\hat{\mathbb{B}}_{k_1k_2\cdots k_p} = \hat{\mathbb{B}}_{k_1k_2\cdots k_p} + n$ *for any integer* n *with* $ord_2(n) > k_p$.

The dual (for $\mathbb{A}$, instead of $\mathbb{B}$) of the above two lemmas can be easily stated and the proofs follow in similar fashion. We state the following lemma which is the dual of Lemma 7.4.3; we omit the proof.

Lemma 7.4.5. *Let* $\mathbb{A} = SFS\{\varepsilon_i 2^i : i \geq 0, i \text{ odd}\}$ *and for an odd integer* $k > 0$ *let* $\mathbb{A}_k = SFS\{\varepsilon_i 2^i : i \geq 0, i \text{ odd}, i \neq k\}$. *For* $\mathbb{C} \in \mathscr{C}(\mathbb{A})$ *let* $\hat{\mathbb{A}}_k = \mathbb{A}_k \oplus \mathbb{C}$. *Then* $\hat{\mathbb{A}}_k = \hat{\mathbb{A}}_k + n$ *for any integer* n *with* $ord_2(n) > k$.

The next lemma is essentially the necessary part of the Finite Extension Theorem and its proof incorporates the basics of the sufficiency proof of Theorem 7.4.2.

Lemma 7.4.6. *Let* $\mathbb{C} \in \mathscr{C}(\mathbb{A})$, *and let* n *be an integer such that* $ord_2(n)$ *is odd. Then there exists a complementing set* $\mathbb{E}$ *of* $\mathbb{C}$ *in* $\mathbb{Z}$ ($\mathbb{E} \in \mathscr{C}(\mathbb{C})$) *such that* $n \in \mathbb{E}$.

Proof. Let $\mathbb{C} \in \mathscr{C}(\mathbb{A})$, and let n be an integer such that $ord_2(n) = k$ where k is odd. Then $n = \varepsilon_k 2^k + l$ for some integer l with $ord_2(l) > k$. We let $\mathbb{A}_k = SFS\{\varepsilon_i 2^i : i > 0, i \text{ odd}, i \neq k\}$ and $\hat{\mathbb{A}}_k = \mathbb{A}_k \oplus \mathbb{C}$. Then

$$\mathbb{A} = (\mathbb{A}_k + \varepsilon_k 2^k) \cup \mathbb{A}_k \quad (disj).$$

We let

$$\mathbb{E} = (\mathbb{A}_k + \varepsilon_k 2^k + l) \cup \mathbb{A}_k \quad (disj).$$

Then 0 belongs to the set $\mathbb{A}_k$, and n belongs to the set $\mathbb{A}_k + \varepsilon_k 2^k + l$; therefore the set $\mathbb{E}$ contains both 0 and n. Also from Lemma 7.4.5 follows

$$\mathbb{E} \oplus \mathbb{C} = (\hat{\mathbb{A}}_k + \varepsilon_k 2^k + l) \cup \hat{\mathbb{A}}_k = (\hat{\mathbb{A}}_k + \varepsilon_k 2^k) \cup \hat{\mathbb{A}}_k = \mathbb{Z}.$$

$\square$

For the proof of the Finite Extension Theorem it will be convenient to use the following notation:
As noted above, with the choice of $\pm$'s prescribed by the sequence $\{\varepsilon_i : i \geq 0\}$ that we have made, the $SFS\{\varepsilon_i 2^i : i \geq 0\}$ equals $\mathbb{Z}$. In other words, every integer $n \in \mathbb{Z}$ has a unique representation as a finite sum of integers of the form $\varepsilon_i 2^i$ for $i \geq 0$ (we write $n = \Sigma_{1 \leq i \leq p} \varepsilon_{n_i} 2^{n_i}$). For notational simplification we suppress the underlying ε_i's, order the indices $0 \leq n_1 < n_2 < \cdots < n_p$ and write this representation of n as $n = (n_1, n_2, \ldots, n_p)$. It is interesting to observe that in this notation if $-n = (n_1', n_2', \ldots, n_q')$ then $n_1 = n_1'$; also $ord_2(n) = n_1$, and 0 has the null representation (sum over the null set) with $ord_2(0) = \infty$.

Proof (Finite Extension Theorem). Let $\mathbb{C} \in \mathscr{C}(\mathbb{A})$. According to Lemma 7.4.6, if for some integer $n \in \mathbb{Z}$, $ord_2(n)$ is odd, then there exists a set $\mathbb{E} \in \mathscr{C}(\mathbb{C})$ with $n \in \mathbb{E}$. Since $(\mathbb{E} - \mathbb{E}) \cap (\mathbb{C} - \mathbb{C}) = \{0\}$ we conclude that $ord_2(n)$ is even for every $n \in \mathbb{C} - \mathbb{C}$. This proves the condition on $ord_2(n)$ is necessary for Theorem 7.4.2.

To prove the sufficiency of the *ord* condition for the theorem we consider $ord_2(n)$ for every $n \in \mathbb{F}$. Let $0 \leq r_1 < r_2 < \cdots < r_p$ be the numbers, listed in increasing order, that represent the ord_2 of the members of $\mathbb{F}$. We note that for every j, $1 \leq j \leq p - 1$, r_j is even, and $r_p = ord_2(0) = \infty$. Next we decompose the set $\mathbb{B} = SFS\{\varepsilon_i 2^i \mid i \geq 0,\ i\ even\}$ stepwise as follows:

$$\mathbb{B} = (\mathbb{B}_{r_1} + \varepsilon_{r_1} 2^{r_1}) \cup \mathbb{B}_{r_1}$$

$$= (\mathbb{B}_{r_1} + \varepsilon_{r_1} 2^{r_1}) \cup (\mathbb{B}_{r_1 r_2} + \varepsilon_{r_2} 2^{r_2}) \cup \mathbb{B}_{r_1 r_2}$$

$$\vdots$$

$$= (\mathbb{B}_{r_1} + \varepsilon_{r_1} 2^{r_1}) \cup (\mathbb{B}_{r_1 r_2} + \varepsilon_{r_2} 2^{r_2}) \cup \cdots$$

$$\cdots \cup (\mathbb{B}_{r_1 r_2 \cdots r_{p-1}} + \varepsilon_{r_{p-1}} 2^{r_{p-1}}) \cup \mathbb{B}_{r_1 r_2 \cdots r_{p-1}}. \tag{7.7}$$

We recall that $\mathbb{B}_{r_1 r_2 \cdots r_j} = SFS\{\varepsilon_i 2^i : i \geq 0,\ i\ even,\ i \neq r_1, r_2, \ldots, r_j\}$ for $1 \leq j < p$. We also use the notation $\hat{\mathbb{E}} = \mathbb{E} \oplus \mathbb{C}$ for any $\mathbb{E} \subset \mathbb{Z}$. At this point we note that for each j, $1 \leq j < p$, there are two possible cases:

Case 1: *There is exactly one $n \in \mathbb{F}$ such that $ord_2(n) = r_j$.*
In this case we write $n = \varepsilon_{r_j} 2^{r_j} + l$ where $ord_2(l) > r_j$. Then we replace the set $\mathbb{B}_{r_1 r_2 \cdots r_j} + \varepsilon_{r_j} 2^{r_j}$ in Eq. (7.7) by $\mathbb{B}_{r_1 r_2 \cdots r_j} + \varepsilon_{r_j} 2^{r_j} + l$, and using Lemma 7.4.4 we note that $\hat{\mathbb{B}}_{r_1 r_2 \cdots r_j} + \varepsilon_{r_j} 2^{r_j} = \hat{\mathbb{B}}_{r_1 r_2 \cdots r_j} + \varepsilon_{r_j} 2^{r_j} + l$. Thus, in this case, the set $\mathbb{B}$ in Eq. (7.7) gets replaced by another set $\mathbb{B}'$ with the property that $n \in \mathbb{B}'$ and $\mathbb{B}' \in \mathscr{C}(\mathbb{A})$.
Next we consider the second case.

Case 2: *There is more than one member of $\mathbb{F}$ with ord_2 equal to r_j.*
We let $\mathbb{F}' = \{n \in \mathbb{F} : ord_2(n) = r_j\}$ and consider the term $\mathbb{B}_{r_1 r_2 \cdots r_j} + \varepsilon_{r_j} 2^{r_j}$ in (7.7). We recall the representation discussed above for every $n \in \mathbb{F}'$. In that notation every $n \in \mathbb{F}'$ has a representation as a finite-tuple of nonnegative and increasing integers, where the first entry is r_j. Let us denote by $(r_j, t_1, t_2, \ldots, t_k)$ the largest initial segment common to all the representations of the members of $\mathbb{F}'$.
We let $m = \varepsilon_{r_j} 2^{r_j} + \varepsilon_{t_1} 2^{t_1} + \varepsilon_{t_2} 2^{t_2} + \cdots + \varepsilon_{t_k} 2^{t_k}$, replace the set $\mathbb{B}_{r_1 r_2 \cdots r_j} + \varepsilon_{r_j} 2^{r_j}$ in (7.7) by $\mathbb{B}_{r_1 r_2 \cdots r_j} + m$, and once more, using Lemma 7.4.4, we note that $\hat{\mathbb{B}}_{r_1 r_2 \cdots r_j} + \varepsilon_{r_j} 2^{r_j} = \hat{\mathbb{B}}_{r_1 r_2 \cdots r_j} + m$. Next we let $\mathbb{G} = \{n : n = f - m,\ f \in \mathbb{F}'\}$, and consider the $ord_2(n)$ for every $n \in \mathbb{G}$. Let $s_1 < s_2 < \cdots < s_q$ be the numbers, listed in increasing order, that represent the ord_2 of the members of $\mathbb{G}$. We note that $r_j < s_1$, and for every i, $1 \leq i \leq q - 1$, s_i is even. Now we repeat the above argument and decompose the set $\mathbb{B}_{r_1 r_2 \cdots r_j} + m$ stepwise as follows:

$$\mathbb{B}_{r_1 r_2 \cdots r_j} + m =$$

$$= (\mathbb{B}_{r_1 \cdots r_j s_1} + m + \varepsilon_{s_1} 2^{s_1}) \cup (\mathbb{B}_{r_1 \cdots r_j s_1} + m)$$

$$= (\mathbb{B}_{r_1 \cdots r_j s_1} + m + \varepsilon_{s_1} 2^{s_1}) \cup (\mathbb{B}_{r_1 \cdots r_j s_1 s_2} + m + \varepsilon_{s_2} 2^{s_2})$$

$$\cup (\mathbb{B}_{r_1 \cdots r_j s_1 s_2} + m)$$

$$\vdots \tag{7.8}$$

$$= (\mathbb{B}_{r_1\cdots r_j s_1} + m + \varepsilon_{s_1} 2^{s_1}) \cup (\mathbb{B}_{r_1\cdots r_j s_1 s_2} + m + \varepsilon_{s_2} 2^{s_2}) \cup \cdots$$

$$\cdots \cup (\mathbb{B}_{r_1\cdots r_j s_1 s_2\cdots s_{q-1}} + m + \varepsilon_{s_{q-1}} 2^{s_q-1}) \cup (\mathbb{B}_{r_1\cdots r_j s_1 s_2\cdots s_{q-1}} + m).$$

Next we shall consider the last term in (7.8), and possibly by altering it ensure that it contains the members of $\mathbb{G}$ that have ord_2 equal to s_q. We note that s_q may be odd. Let $\mathbb{G}' = \{n \in \mathbb{G} : ord_2(n) = s_q\}$, and as before we denote by $(s_q, t_1', t_2', \ldots, t_{k'}')$ the largest initial segment common to all the representations of the members of $\mathbb{G}'$. Let $m' = \varepsilon_{s_q} 2^{s_q} + \varepsilon_{t_1'} 2^{t_1'} + \varepsilon_{t_2'} 2^{t_2'} + \cdots + \varepsilon_{t_{k'}'} 2^{t_{k'}'}$ and replace the set $\mathbb{B}_{r_1\cdots r_j s_1\cdots s_{q-1}} + m$, the last term in (7.8), by $\mathbb{B}_{r_1\cdots r_j s_1\cdots s_{q-1}} + m + m'$; once more using Lemma 7.4.4 $\hat{\mathbb{B}}_{r_1\cdots r_j s_1\cdots s_{q-1}} + m = \hat{\mathbb{B}}_{r_1\cdots r_j s_1\cdots s_{q-1}} + m + m'$.

Next we let $\mathbb{H} = \{n : n = g - m, g \in \mathbb{G}'\}$ and consider $ord_2(n)$ for every $n \in \mathbb{H}$. Let $s_1' < s_2' < \cdots < s_{q'}'$ be the numbers, listed in increasing order, that represent the ord_2 of the members of $\mathbb{H}$. We note that $s_{q-1} < s_1'$, and for every i, $1 \le i \le q' - 1$, s_i' is even. Again we repeat the above argument and decompose the set $\mathbb{B}_{r_1\cdots r_j s_1\cdots s_{q-1}} + m + m'$ as

$$\mathbb{B}_{r_1\cdots r_j s_1\cdots s_{q-1}} + m + m' =$$

$$(\mathbb{B}_{r_1\cdots r_j s_1\cdots s_{q-1} s_1'} + m + m' + \varepsilon_{s_1'} 2^{s_1'}) \cup (\mathbb{B}_{r_1\cdots r_j s_1\cdots s_{q-1} s_1' s_2'} + m + m' + \varepsilon_{s_2'} 2^{s_2'}) \cup \cdots$$

$$\cup (\mathbb{B}_{r_1\cdots r_j s_1\cdots s_{q-1} s_1'\cdots s_{q'-1}'} + m + m' + \varepsilon_{s_{q'-1}'} 2^{s_{q'-1}'})$$

$$\cup (\mathbb{B}_{r_1\cdots r_j s_1\cdots s_{q-1} s_1'\cdots s_{q'-1}'} + m + m'). \tag{7.9}$$

We replace the last term of (7.8) by the decomposition (7.9). Next we work on the last term of the decomposed (7.8) by considering the elements of $\mathbb{H}$ that have ord_2 equal to $s_{q'}'$. We repeat this process, and after a finite number of steps we succeed replacing the set $\mathbb{B}_{r_1\cdots r_j} + \varepsilon_{r_j} 2^{r_j}$ in (7.7) by a decomposition where each term will satisfy either one of the two cases mentioned above.

We keep repeating the above argument, and this way we continue decomposing the sets that correspond to Case 2 into finer and finer subsets and replace them by their appropriate translates. Eventually, in a finite number of steps we succeed in replacing the set $\mathbb{B}$ by a union of sets where each one is reduced to Case 1.   □

The above theorem can be generalized to any infinite subsets $\mathbb{A}$ and $\mathbb{B}$ with the property that $\mathbb{A} \oplus \mathbb{B} = \mathbb{N}$. The notion of ord_2 of a number needs to be generalized further and the notation becomes more complicated; however, all the arguments and proofs used here carry over without any difficulty.

7.5   Complementing Sets of $\mathbb{A}$ and the 2-Adic Integers

In this section we continue our investigation of complementing sets of $\mathbb{A}$ in $\mathbb{Z}$ ($\mathscr{C}(\mathbb{A})$) where the set $\mathbb{A} = SFS\,\{2^{2i+1} : i \geq 0\}$.

We consider the following two conditions for a set of integers.

Conditions 7.5.1 *For a subset $\mathbb{C}$ of integers we consider the following two conditions:*

(i) *For every $c, c' \in \mathbb{C}$ either $c = c'$ or the maximal number i such that 2^i divides $c - c'$ is even.*

(ii) *$\mathbb{C}$ is maximal with respect to (i). That is if $\mathbb{C}'$ satisfies (i) and $\mathbb{C} \subset \mathbb{C}'$ then $\mathbb{C} = \mathbb{C}',$*

From the results in the previous section the following is clear.

Lemma 7.5.2. *A necessary condition for a set of integers $\mathbb{C}$ to be a complementing set in $\mathbb{Z}$ for the set $\mathbb{A}$ is that $\mathbb{C}$ satisfies (i) and (ii) in Conditions 7.5.1.*

Proof. Suppose $\mathbb{A} \oplus \mathbb{C} = \mathbb{Z}$. The Finite Extension Theorem (Theorem 7.4.2) shows that (i) is necessary. If $\mathbb{C}'$ satisfies (i) and $n \in \mathbb{C}' \setminus \mathbb{C}$ then $n = a + c$ for some $a \in \mathbb{A}$ and $c \in \mathbb{C}$. Thus $n - c = a$ is a difference whose $(ord_2(n - c))$ highest integer i so that 2^i divides $n - c$ is even and also odd (since $n - c \in \mathbb{A}$); therefore $n - c$ must be 0. That is, (ii) holds.   □

The conditions on $\mathbb{C}$, however, are clearly not sufficient for $\mathbb{C}$ to be complementing $\mathbb{A}$ in $\mathbb{Z}$. The set $\mathbb{B} = SFS\,\{2^{2i} : i \geq 0\}$ satisfies both conditions, but obviously $\mathbb{B}$ is not a complementing set of $\mathbb{A}$ in $\mathbb{Z}$.

Clearly, $\mathbb{C} = -\mathbb{B}$ is a complementing set of $\mathbb{A}$ in $\mathbb{Z}$, and it is tempting to conjecture that with enough negative integers in $\mathbb{C}$ it will also be a complementing set of $\mathbb{A}$ in $\mathbb{Z}$. To illustrate this point we state the following without proof.

Lemma 7.5.3. *Let*

$$\mathbb{B}_\omega = \Big\{ \sum_{i=0}^{\infty} \varepsilon_i \omega_i 2^{2i} \ : \ \varepsilon_i \in \{0, 1\} \ and \ \varepsilon_i = 1 \ for \ finitely \ many \ i's \Big\}$$

where $\omega \in \{-1, 1\}^{\mathbb{N}}$ and $\omega_i = -1$ for infinitely many i's. Then $\mathbb{A} \oplus \mathbb{B}_\omega = \mathbb{Z}$.

The sets $\mathbb{B}_\omega$ obviously have enough negative integers and can be used to guarantee enough negative integers in a possible complementing set $\mathbb{C}$—that this is part of a sufficient condition for $\mathbb{C}$ to be a complementing set is made precise by the next theorem [41].

Theorem 7.5.4. *Let $\mathbb{C}$ be a subset of $\mathbb{Z}$ containing 0. Suppose $\mathbb{C}$ satisfies both conditions (i) and (ii) of Conditions 7.5.1 above. Then $\mathbb{A} \oplus \mathbb{C} = \mathbb{Z}$ if and only if $\mathbb{C}$ satisfies:*

(iii) *There exists an ω as above in Lemma 7.5.3 such that $\mathbb{A} \oplus \mathbb{C} \supset \mathbb{B}_\omega$.*

We will not prove this theorem. It is tempting to try and extend the condition (iii) on $\mathbb{C}$ to

(iv) There exists a $\mathbb{D} \in \mathscr{C}(\mathbb{A})$ such that $\mathbb{A} \oplus \mathbb{C} \supset \mathbb{D}$.

In Sect. 7.5.2, we show how by viewing everything in the 2-adic integers, condition (iv) on $\mathbb{C}$ above can be seen to be insufficient to guarantee that $\mathbb{C}$ is a complementing set of $\mathbb{A}$ in $\mathbb{Z}$.

7.5.1 The 2-Adic Integers

In this section we present and discuss some results on the 2-adic integers (see [16, 20]).

Let

$$\mathbb{Z}_2 = \left\{ z = \sum_{i \geq 0} z_i 2^i \; : \; z_i \in \{0, 1\} \right\}$$

denote the completion of $\mathbb{Z}$ in the 2-adic valuation norm (the norm of z is 2^{-i} where i is the index of the first nonzero entry in z). For notational convenience we identify $\mathbb{Z}_2$ with $\{0, 1\}^{\mathbb{N}}$, i.e $z = \sum z_i 2^i \leftrightarrow (z_0, z_1, z_2, \ldots)$; we will also write $\overline{\mathbb{Z}}$ for $\mathbb{Z}_2$, regarding it as the closure of the integers $\mathbb{Z}$ in the 2-adic integers. The positive integers are represented by $n = (z_0, z_1, z_2, \ldots)$ with $z_i = 0$ for all but finitely many i's. The negative integers are represented by $m = (z_0, z_1, z_2, \ldots)$ with $z_i = 1$ for all but finitely many i's.

We consider the ord_2 function (previously defined for positive integers n) extended to $\mathbb{Z}_2$: for $z = (z_0, z_1, z_2, \ldots) \in \mathbb{Z}_2$ define $ord(z) = i$ if $z_i = 1$ and $z_j = 0$ for all $0 \leq j < i$. Recall that for $n \in \mathbb{N}$, $ord(n) = ord_2(n)$ is the highest power of 2 which divides n. The ord function is used in analyzing the distance between two numbers $c, d \in \mathbb{Z}_2$; that is, $ord(c - d) = n$ means that c, d are the same for the first n coordinates $c_i = d_i$ for $0 \leq i \leq n - 1$; the distance between c, d is defined as $2^{-ord(c-d)}$. We will often be concerned with whether the ord is even or odd. Note that $ord(0) = \infty$ and $ord(0)$ is considered both odd and even.

Recalling the two properties in Conditions 7.5.1, a subset $\mathbb{E}$ of $\mathbb{Z}_2$ is said to *have even (odd) differences* if $ord_2(e - e')$ is even (respectively odd) for all $e \neq e'$ in $\mathbb{E}$. A set of integers $\mathbb{C}$ which has even differences is said to be *maximal in $\mathbb{Z}$* if it satisfies (ii) of 7.5.1 (i.e., if $\mathbb{C}' \supset \mathbb{C}$, and $\mathbb{C}'$ is a subset of $\mathbb{Z}$ which has even differences then $\mathbb{C}' = \mathbb{C}$). A subset $\mathbb{E}$ of $\mathbb{Z}_2$ which has even differences is *maximal in $\mathbb{Z}_2$* if any subset of $\mathbb{Z}_2$ containing $\mathbb{E}$ which has even differences coincides with $\mathbb{E}$. We will simply use the term *maximal* when it is clear from the context which definition applies. (Similar definitions can be made for odd differences.)

A set of integers $\mathbb{C}$ is *even complete* if for all $n \geq 1$ and for every $\xi \in \{0, 1\}^n$, $\xi = (\xi_0, \xi_1, \ldots, \xi_{n-1})$, there exists an integer $c = (c_0, c_1, \ldots) \in \mathbb{C} \subset \mathbb{Z}_2$ with $c_{2i} = \xi_i$, $0 \leq i \leq n - 1$. Similarly, a set of integers $\mathbb{D}$ is *odd complete* if for all $n \geq 1$ and for every $\xi = (\xi_0, \xi_1, \ldots, \xi_{n-1}) \in \{0, 1\}^n$ there exists an integer $d = (d_0, d_1, \ldots) \in \mathbb{D} \subset \mathbb{Z}_2$ with $d_{2i+1} = \xi_i, 0 \leq i \leq n - 1$.

Lemma 7.5.5. *Let $\mathbb{C}$ be a set of integers containing 0 which has even differences and is maximal in $\mathbb{Z}$. Then $\mathbb{C}$ is even complete.*

Proof. This is essentially contained in Lemma 1 of [5]. Let $n \geq 1$ be the smallest integer such that there exists $(\xi_0, \ldots, \xi_{n-1})$ and no $c = (c_0, c_1, \ldots) \in \mathbb{C}$ with $c_{2i} = \xi_i, 0 \leq i \leq n - 1$. If $n = 1$ there are two cases depending on the value of ξ_0. If $\xi_0 = 1$ then $\mathbb{C}$ contains no odd integers and 1 may be adjoined to $\mathbb{C}$ and maintain even differences. If $\xi_0 = 0$ it means all integers in $\mathbb{C}$ are odd and the number 4 may be adjoined. If $n > 1$ let $c = (c_0, c_1, \ldots) \in \mathbb{C}$ with $c_{2i} = \xi_i, 0 \leq i \leq n - 2$. Hence $c_{2n-2} \neq \xi_{n-1}$ and $ord(c - c') \neq 2n - 2$ for all $c' \in \mathbb{C}$. The number $c + 2^{2n-2}$ may then be adjoined to $\mathbb{C}$ as $ord(c + 2^{2n-2} - c') = min(ord(c - c'), 2n - 2)$.   □

The two conditions, even differences and even completeness, are not enough to make a set a complementing set of $\mathbb{A}$ or even maximal. Consider the set $-\mathbb{B}$ and remove from it the number -1 (i.e., $\mathbb{B}' = -\mathbb{B} \setminus \{-1\}$). Then $\mathbb{B}'$ remains even complete, but is not maximal with respect to even differences, and it is not a complementing set of $\mathbb{A}$. Observe, however, that -1 is in the closure of this set.

The following two lemmas appear in [16] in a more general form and are variations of Lemma 3 in [46].

Lemma 7.5.6. *Let $\mathbb{C}$ be a set of integers containing 0 which has even differences and is even complete. Then the closure $\overline{\mathbb{C}}$ has even differences and is maximal with respect to even differences in $\mathbb{Z}_2$. That is if $\mathbb{C}' \supset \overline{\mathbb{C}}$ and $\mathbb{C}'$ has even differences then $\mathbb{C}' = \overline{\mathbb{C}}$.*

The corresponding result for odd differences in place of even differences also holds.

Proof. Let $z, z' \in \overline{\mathbb{C}}$ with $ord(z - z') = n$. Choose $c, c' \in \mathbb{C}$ with $ord(c - z) > n$ and $ord(c' - z') > n$. Hence $ord(c - c') = n$ and so is even; i.e. $\overline{\mathbb{C}}$ has even differences.

To show the maximality in $\mathbb{Z}_2$, suppose $x \in \mathbb{C}' \supset \mathbb{C}$. Then if $\mathbb{C}'$ has even differences, then x satisfies $ord(z - x)$ is even for all $z \in \overline{\mathbb{C}}$. Put $\xi_i = x_{2i}, i \geq 0$. Then for each $n \geq 0$, by the definition of even complete, there must be a $c(n) \in \mathbb{C}$ with $(c(n))_{2i} = x_{2i}$ for $0 \leq i < n$. By the even differences of $ord(c(n) - x)$ it follows that $(c(n))_j = x_j, 0 \leq j \leq 2n$. Therefore the sequence $c(n) \in \mathbb{C}$ converges to x and so x is in the closure of $\mathbb{C}$.   □

Lemma 7.5.7. *Let $\mathbb{C}$ be a set of integers containing 0 which has even differences and is even complete. Let $\mathbb{E}$ be a set of integers containing 0 which has odd differences and is odd complete. Then*

$$\overline{\mathbb{C}} \oplus \mathbb{E} = \overline{\mathbb{C}} \oplus \overline{\mathbb{E}} = \mathbb{Z}_2.$$

Proof. The sumset $\mathbb{C} + \mathbb{E}$ is direct: if $c + d = c' + d'$ for $c, c' \in \mathbb{C}$ and $d, d' \in \mathbb{E}$, then $c - c' = d - d'$. Hence this difference has both even and odd *ord*, and so must be 0. The denseness of $\mathbb{C} \oplus \mathbb{E}$ is similar to the reasoning in the previous proof. The equality of the closure of the sum with the sum of the closures is straightforward from the odd/even differences. □

Lemma 7.5.8 supplies a converse to Lemma 7.5.5.

Lemma 7.5.8. *Let $\mathbb{C}$ be a set of integers containing 0 which has even differences and is even complete. Then $\mathbb{C}' = \overline{\mathbb{C}} \cap \mathbb{Z}$ has even differences and is maximal in $\mathbb{Z}$.*

Remark 7.5.9. Lemma 7.5.7 explains how a set $\mathbb{C}$ can satisfy Conditions 7.5.1 yet not be a complementing set of $\mathbb{A}$ in $\mathbb{Z}$. For any integer n which is not in $\mathbb{A} \oplus \mathbb{C}$ there must be $\hat{a} \in \overline{\mathbb{A}} \backslash \mathbb{A}$ and $\hat{c} \in \overline{\mathbb{C}} \backslash \mathbb{C}$, so that $\hat{a} + \hat{c} = n$. Observe that any $\hat{a} = (a_0, a_1, \ldots) \in \overline{\mathbb{A}}$ has 1's only in odd locations, i.e. $a_{2i} = 0$ for all i, and $\hat{a} \in \overline{\mathbb{A}} \backslash \mathbb{A}$ means $a_{2i+1} = 1$ for infinitely many i. As an illustration, consider the set $\mathbb{B}$ which satisfies Conditions 7.5.1 but is not a complement. The numbers $-1/3 = (1, 0, 1, 0, \overline{1, 0}) \in \overline{\mathbb{B}}$ and $-2/3 = (0, 1, 0, 1, \overline{0, 1}) \in \overline{\mathbb{A}}$ and so -1 is not in $\mathbb{A} \oplus \mathbb{B}$. Note that both $\mathbb{A}$ and $\mathbb{B}$ are positive and so obviously the sum contains no negative integers.

Theorem 7.5.10. *Let $\mathbb{C}$ be a subset of $\mathbb{Z}$ containing 0 and satisfying (i) and (ii) in Conditions 7.5.1; that is:*

(i) $\mathbb{C}$ has even differences,
(ii) $\mathbb{C}$ is maximal in $\mathbb{Z}$ with respect to containing even differences.
Then $\mathbb{C} \in \mathscr{C}(\mathbb{A})$ if and only if $\mathbb{C}$ satisfies:
(v) for any $\hat{c} = (c_0, c_1, \ldots) \in \overline{\mathbb{C}} \backslash \mathbb{C}$ there are infinitely many i so that $c_{2i} = 0$.

Proof. Assume $\mathbb{C}$ satisfies the two conditions (i) and (ii) from Conditions 7.5.1. First we note that any $\hat{c} \in \overline{\mathbb{C}} \backslash \mathbb{C}$ cannot be an integer, since otherwise $\mathbb{C}' = \hat{c} \cup \mathbb{C}$ has even differences (Lemma 7.5.6) and would contradict (ii), the maximality of $\mathbb{C}$. Then we may assume $\hat{c} \in \overline{\mathbb{C}} \backslash \mathbb{C}$ is not an integer. Therefore there are infinitely many i with $c_i = 0$ and infinitely many i with $c_i = 1$.

Suppose $\hat{c}$ has only finitely many i such that $c_{2i} = 0$. Then there is an $n > 0$ such that if $i \geq n$ and $c_i = 0$ then i must be odd. Denote the collection of these i as I. We define $\hat{a} \in \overline{\mathbb{A}} \backslash \mathbb{A}$ by $a_i = 1$ for all indices $i \in I$ and nowhere else. Then $\hat{a} + \hat{c}$ is a negative integer and $\mathbb{C}$ cannot be a complementing set.

Suppose there are infinitely many i with $c_{2i} = 0$. Denote this set of i by I. We claim that there is no $\hat{a} \in \overline{\mathbb{A}}$ with $\hat{a} + \hat{c}$ an integer. In order for $\hat{a} + \hat{c}$ to be a negative integer the sum $\hat{a} + \hat{c}$ must have a 1 in all but a finite number of the coordinates $i \in I$. Since these are even there must have been a carry from a lower indexed coordinate. Consider $i < j$ and $i, j \in I$ such that there is no $k \in I$ with $i < k < j$. There can be no carry from the $2i$ coordinate. Hence to get a carry into the $2j$ coordinate there must be an odd coordinate $2k + 1$, with $2i < 2k + 1 < 2j$, which starts the carry. But then the $2k + 1$ coordinate of $\hat{a} + \hat{c}$ must be 0 and the sum cannot be an integer. A similar argument shows that the sum cannot be a positive integer. □

7.5.2   Condition (iv) Is Not Enough to Be Complementing

We are now in position to show by example that condition (iv) (after Theorem 7.5.4) is not sufficient to guarantee that $\mathbb{C}$ is a complementing set of $\mathbb{A} = SFS\{2^{2i+1} : i \geq 0\}$; i.e. we will construct two subsets of the integers $\mathbb{C}$ and $\mathbb{D}$ which both satisfy Conditions 7.5.1. The set $\mathbb{C}$ will not be a complementing set of $\mathbb{A}$, but the set $\mathbb{D}$ will be a complementing set in $\mathbb{Z}$ of $\mathbb{A}$ and $\mathbb{A} \oplus \mathbb{C} \supset \mathbb{D}$. These sets are a variation of Example 4.2 appearing in [41] and are both built from the same general construction.

We define

$$\mathbb{C}_{\mathbf{p}} = \bigcup_{i \geq 1} \{p_i - 2^{2i}\mathbb{B}\} = \bigcup_{i \geq 1} \{p_i - 2^{2i}b \mid b \in \mathbb{B}\}$$

where the set $\mathbb{B} = SFS\{2^{2i} : i \geq 0\}$ as defined earlier in Sect. 7.5, and $\mathbf{p} = \{p_k : k \geq 1\}$ is a sequence of integers satisfying:

(a) $p_1 = 0$,
(b) $1 \leq p_i < 2^{2i}$ and p_i is an odd integer for each $i \geq 2$,
(c) $ord_2(p_i - p_{i+1}) = 2(i - 1)$.

The utility of this construction is given in the following lemma.

Lemma 7.5.11. *For the sequence* $\mathbf{p} = (p_1, p_2, \ldots)$ *and the set* $\mathbb{C}_{\mathbf{p}}$, *as defined above, the following properties hold:*

1. p_i *converge to some* $\hat{p}$ *in* $\mathbb{Z}_2$,
2. $\mathbb{C}_{\mathbf{p}}$ *has even differences,*
3. $\mathbb{C}_{\mathbf{p}}$ *is even complete,*
4. $\mathbb{C}_{\mathbf{p}}$ *is maximal in* $\mathbb{Z}$ *if and only if* $\hat{p}$ *is not an integer,*
5. $\mathbb{C}_{\mathbf{p}}$ *is a complementing set if and only if*
   $\hat{p} \in \overline{\mathbb{Z}} \setminus \mathbb{Z}$ *is not an integer, and there does not exist an* $a \in \overline{\mathbb{A}}$ *with* $a + \hat{p}$ *an integer.*

We first present a few examples including the two sets for the counterexample, and we end the section with the proof of the above lemma. (Because of property 1 in Lemma 7.5.11 we rewrite the notation for $\mathbb{C}_{\mathbf{p}}$ below as $\mathbb{C}_{\hat{p}}$.)

Example 7.5.12. $\mathbb{C}_{-1/3} : p_k = \displaystyle\sum_{i=0}^{k-2} 2^{2i}$ for $k \geq 2$.

This example appears in [41]. Some of the terms in their 2-adic descriptions follow, $p_1 = (0, \bar{0})$, $p_2 = (1, 0, \bar{0})$, $p_3 = (1, 0, 1, 0, \bar{0})$, $p_4 = (1, 0, 1, 0, 1, 0, \bar{0})$. It is easy to see that the limit is $\hat{p} = (1, 0, \overline{1, 0}) = -1/3$. Since $-2/3 = (0, 1, \overline{0, 1}) \in \overline{\mathbb{A}}$ property 5 implies that the set $\mathbb{C}_{-1/3}$ is not a complementing set of $\mathbb{A}$.

Example 7.5.13. $\mathbb{C}_{-1} : p_k = \sum_{i=0}^{k-2} \left(2^{2i} + 2^{2i+1} \right)$ for $k \geq 2$.

These terms are $p_1 = 0 = (0, \overline{0})$, $p_2 = (1, 1, 0, \overline{0})$, $p_3 = (1, 1, 1, 1, 0, \overline{0})$, and $p_4 = (1, 1, 1, 1, 1, 1, 0, \overline{0})$. The limit of this sequence is $\hat{p} = (1, \overline{1}) = -1$. In this case $\overline{\mathbb{C}}_{-1}$ is not maximal and also not a complementing set for $\mathbb{A}$.

Example 7.5.14. $\mathbb{C}_{-7/15} : p_k = \sum_{i=0}^{k-2} 3^{(i+1) \bmod 2} \cdot 2^{2i}$ for $k \geq 2$.

A few representations of these terms are $p_1 = 0$, $p_2 = (1, 1, 0, \overline{0})$, $p_3 = (1, 1, 1, 0, \overline{0})$, $p_4 = (1, 1, 1, 0, 1, 1, 0, \overline{0})$ and $p_5 = (1, 1, 1, 0, 1, 1, 1, 0, \overline{0})$. The limit of the sequence is $\hat{p} = (1, 1, 1, 0, \overline{1, 1, 1, 0}) = -7/15$. The set $\mathbb{C}_{-7/15}$ is not a complementing set because $-8/15 = (0, 0, 0, 1, \overline{0, 0, 0, 1}) \in \overline{\mathbb{A}}$.

Example 7.5.15. $\mathbb{C}_{-9/15} : p_k = \sum_{i=0}^{k-2} \left(3^{(i+1) \bmod 2} \cdot 2^{2i} + \left((i+1) \bmod 2 \right) \cdot 2^{2i+1} \right)$ for $k \geq 2$.

A few of these terms are $p_1 = 0$, $p_2 = (1, 0, 1, 0, \overline{0})$, $p_3 = (1, 0, 0, 1, 0, \overline{0})$, $p_4 = (1, 0, 0, 1, 1, 0, 1, 0, \overline{0})$ and $p_5 = (1, 0, 0, 1, 1, 0, 0, 1, 0, \overline{0})$. The limit of this sequence in the 2-adics is $\hat{p} = (1, 0, 0, 1, \overline{1, 0, 0, 1}) = -9/15$. From the above lemma as well as Theorem 7.5.10 it follows that $\mathbb{C}_{-9/15}$ is a complementing set of $\mathbb{A}$.

Counterexample. The sets $\mathbb{C}_{-7/15}$ and $\mathbb{C}_{-9/15}$ form the promised counterexample. To see that $\mathbb{A} \oplus \mathbb{C}_{-7/15} \supset \mathbb{C}_{-9/15}$ simply observe that the difference of the p_k for $\mathbb{C}_{-9/15}$ and $\mathbb{C}_{-7/15}$ is $q_k = \sum_{i=0}^{k-2} \left((i+1) \bmod 2 \right) \cdot 2^{2i+1} \in \mathbb{A}$.

Proof (Lemma 7.5.11). Properties 1 and 2 are clear from their definitions.

To see property 3 we begin by observing that $-\mathbb{B}$ is even complete. Hence for all $(\xi_0, \ldots, \xi_{n-1})$ with $\xi_1 = 0 = (p_1)_0$ there is a $c \in \{p_1 - 2^2 \cdot \mathbb{B}\}$ with $c_{2i} = \xi_i$ for $0 < i \leq n - 1$.

Next we look at all $(\xi_0, \ldots, \xi_{n-1})$ with $\xi_0 = 1 = (p_2)_0$ and $\xi_2 = (p_2)_2$. Since $1 \leq p_2 < 2^4$ it is clear that for each of these patterns there is a $c \in \{p_2 - 2^4 \cdot \mathbb{B}\}$ with $c_{2i} = \xi_i, 0 < i \leq n - 1$.

We do not know what $(p_2)_2$ is (either 0 or 1), but we have by assumption $ord(p_2 - p_3) = 2^{2 \cdot (2-1)} = 2^2$. This means that $(p_3)_0 = (p_2)_0$ and $(p_3)_2 = ((p_2)_2 + 1) \bmod 2$. Hence for each $(\xi_0, \ldots, \xi_{n-1})$ with $\xi_1 = (p_3)_0$, $\xi_2 = (p_3)_2$ and $\xi_3 = (p_3)_4$ there is a $c \in \{p_3 - 2^6 \cdot \mathbb{B}\}$ with $c_{2i} = \xi_i$ for $0 < i \leq n - 1$.

It is easy to see that the proof of even completeness continues by induction.

The proof of property 4 follows from Lemmas 7.5.6 and 7.5.8. First we observe that if $\hat{c} \in \overline{\mathbb{C}}_{\hat{p}} \backslash \mathbb{C}_{\hat{p}}$ then either $\hat{c} \in \{p_i - 2^{2i} \mathbb{B}\}$ for some i or $\hat{c} = \hat{p}$. It is clear that $\overline{p_i - 2^{2i} \mathbb{B}}$ contains no integers so the only possible additional integer in $\overline{\mathbb{C}}_{\hat{p}}$ can be $\hat{p}$.

Finally, property 5 follows by Remark 7.5.9. $\qquad \square$

7.6   Examples: Non-isomorphic Transformations

If $\mathbb{A}$ and $\mathbb{B}$ are both infinite and $\mathbb{A} \oplus \mathbb{B} = \mathbb{N}$ then there is an obvious symmetry (duality) between them. Assuming $1 \in \mathbb{B}$, it also follows from Theorem 7.1.2 that there is an integer $k > 1$ such that k divides every element in $\mathbb{A}$.

It follows that since we can construct a transformation associated to one set $\mathbb{A}$ (we'll call it $T_{\mathbb{A}}$), we can also construct a transformation (call it $T_{\mathbb{B}}$) associated to the other set $\mathbb{B}$ for the complementing pair in $\mathbb{N}$. We specialize to the two sequences in the First Basic Example; namely $\mathbb{A} = SFS\{2^{2i+1} : i \geq 0\}$ and $\mathbb{B} = SFS\{2^{2i} : i \geq 0\}$. Despite the fact that the two sequences are so similar ($\mathbb{A} = 2\mathbb{B}$) the two transformations associated to the complementing pairs in $\mathbb{N}$ are not isomorphic.

In order to see this non-isomorphism we will construct measures on the 2-adic integers where both transformations naturally reside. We recall that the two-adic integers form an abelian group where addition is defined coordinatewise with "carry to the right." In this section, in addition to the Haar measure μ on $\mathbb{Z}_2$ (which is the same as the product measure $\prod_{i=0}^{\infty}(1/2, 1/2)$) we define two mutually singular measures (and singular with respect to μ as well). We define a single point map T on the 2-adics which will be measure-preserving for each of these measures as well as for μ:

$$T(x_0, x_1, x_2, \ldots) = (x_0, x_1, x_2, \ldots) + (1, 0, 0, \ldots)$$

with the sum having carry to the right.

For technical reasons we remove from our space $\overline{0} = (0, 0, 0, \ldots)$ and its orbit under T, as well as $\overline{1} = (1, 1, 1, \ldots)$ and its preimages (a countable set of points). Call the resulting space X. The carriers of the measures to be defined will be disjoint and connected to the sets $\mathbb{A}$ and $\mathbb{B}$ above. Thus the single map T will be viewed as three different maps $T_{\mathbb{A}}$, $T_{\mathbb{B}}$, and T with respect to the three different measures. We will show that the two infinite measure-preserving maps $T_{\mathbb{A}}$ and $T_{\mathbb{B}}$ are non-isomorphic.

7.6.1   Two Non-isomorphic Transformations

In the space $\mathbb{Z}_2$ we consider two subsets W_e and W_o;

$$W_e = \{x \in X : x_{2i} = 0 \text{ for all } i\},$$

$$W_o = \{x \in X : x_{2i+1} = 0 \text{ for all } i\}.$$

Essentially, W_e (respectively W_o) "lives" on the odd (even) coordinates and is a subset of μ-measure 0 in $\mathbb{Z}_2$. On W_e and W_o we define two product measures. Consider the discrete probability measures $P = (1/2, 1/2)$ and $Q = (1, 0)$ on the two point set $\{0, 1\}$. On W_e we define the infinite product measure

$$\mu_e = \prod_{i=0}^{\infty} \eta_i, \quad \eta_i = \begin{cases} P & \text{if } i \text{ is odd,} \\ Q & \text{if } i \text{ is even.} \end{cases}$$

On W_o we consider the infinite product measure

$$\mu_o = \prod_{i=0}^{\infty} \eta_i, \quad \eta_i = \begin{cases} P & \text{if } i \text{ is even,} \\ Q & \text{if } i \text{ is odd.} \end{cases}$$

Even though both of the sets W_e and W_o have μ-measure 0 we can define on each of these sets the induced transformations, T_e and T_o respectively induced by T via the first return map and we use the notation (W_e, μ_e, T_e) and (W_o, μ_o, T_o) to emphasize the sets on which the transformations are operating.

It is easy to check that the measures μ_e on W_e and μ_o on W_o are invariant under the transformations T_e and T_o respectively, and that both (W_e, μ_e, T_e) and (W_o, μ_o, T_o) are measure-theoretically isomorphic to (X, μ, T)—which in turn is isomorphic to the von Neumann transformation (see Sect. 4.1.2) .

We define the sets

$$X_e = \bigcup_{n \in \mathbb{Z}} T^n W_e$$

$$X_o = \bigcup_{n \in \mathbb{Z}} T^n W_o.$$

Lemma 7.6.1. *The sets X_o and X_e are invariant under T and satisfy the following:*

1. $X_o = \{x \in X : \text{ for some integer } k \geq 0, \; x_{2i+1} = 0 \text{ for all } i \geq k\}$,
2. $X_e = \{x \in X : \text{ for some integer } k \geq 0, \; x_{2i} = 0 \text{ for all } i \geq k\}$.

Proof. Since we removed points whose tails ended in all 0's or all 1's adding (or subtracting) $(1, 0, 0, \ldots)$ to any point will change only a finite number of coordinates of the point. □

Remark 7.6.2. Each $n \in \mathbb{N}$ may be viewed as members of $\mathbb{Z}_2$ whose expansions end in a tail of 0's. Hence the action of T^n (respectively T^{-n}) may also be viewed as addition (respectively subtraction) by n in X. This means that for $b \in \mathbb{B}$ the action of T^b adds 1 to distinct even coordinates, and for $a \in \mathbb{A}$ the action of T^a adds 1 to distinct odd coordinates.

The next two lemmas follow immediately from the above remark, and we state them without proof.

Lemma 7.6.3. *The sets W_e and W_o satisfy:*

1. *For distinct $b, b' \in \mathbb{B}$, $T^b W_e \cap T^{b'} W_e = \emptyset$.*
2. *For distinct $a, a' \in \mathbb{A}$, $T^a W_o \cap T^{a'} W_o = \emptyset$.*

Lemma 7.6.4. *The sets X_e, W_e, X_o and W_o are related as follows:*

1. $X_e = \bigcup_{b \in \mathbb{B}} T^b W_e (disj).$
2. $X_o = \bigcup_{a \in \mathbb{A}} T^a W_o (disj).$

The measure μ_e (respectively μ_o) on W_e (respectively W_o) can be pushed forward to $T^b W_e$ (respectively $T^a W_o$) in the obvious manner. This then allows us to extend the measures μ_e and μ_o to X_e and X_o respectively:

$$\overline{\mu}_e(E) = \sum_{b \in \mathbb{B}} \mu_e(T^{-b}(E \cap T^b W_e)),$$

$$\overline{\mu}_o(E) = \sum_{a \in \mathbb{A}} \mu_o(T^{-a}(E \cap T^a W_o)).$$

The measures $\overline{\mu}_e$ and $\overline{\mu}_o$ are thus infinite and invariant under T_e and T_o respectively. In view of the fact that the induced transformation (W_o, μ_o, T_o) is isomorphic to (X, μ, T), it is clear that the transformation $T_\mathbb{A} = (X_o, \overline{\mu}_o, T_o)$ is isomorphic to our First Basic Example discussed in Chap. 4. We can show also that the transformation $T_\mathbb{B} = (X_e, \overline{\mu}_e, T_e)$ is isomorphic to a transformation built on top of (W_e, μ_e, T_e) by using the sequence $\mathbb{B}$ instead of $\mathbb{A}$ to determine the number of floors to put on above the base W_e. We give a more detailed discussion of these constructions in Sect. 7.7 valid for any infinite complementing pairs $\mathbb{A}$ and $\mathbb{B}$ of a direct sum decomposition of $\mathbb{N}$ (and not just for the special complementing pair $\mathbb{A}$ and $\mathbb{B}$ discussed in this section).

We proceed to show the non-isomorphism of $T_\mathbb{A}$ and $T_\mathbb{B}$.

Lemma 7.6.5. *Let $n > 0$ be a positive integer. Then either $T^n W_e \cap W_e = \emptyset$ or $T^n W_o \cap W_o = \emptyset$.*

Proof. Let $n > 0$ be a positive integer, and suppose that the first time a 1 appears in the dyadic representation of n is in the i-th coordinate. If i is an even integer then $T^n W_e \cap W_e = \emptyset$; otherwise $T^n W_o \cap W_o = \emptyset$. □

Proposition 7.6.6. *The transformations $T_\mathbb{A}$ and $T_\mathbb{B}$ are not isomorphic.*

Proof. Let $\phi : X_e \to X_o$ be an isomorphism between (X_o, T_o) and (X_e, T_e). From Lemma 7.6.5 follows that $T_o^n(\phi W_e \cap W_o) \cap (\phi W_e \cap W_o) = \emptyset$ for any integer n. This says that $\phi W_e \cap W_o$ is a wandering set for T_o on X_o. By a similar argument we conclude that $\phi T_e^i W_e \cap T_o^j W_o$ is a wandering set for any $i, j \in \mathbb{Z}$. We also have $X_o = \cup_{i,j \in \mathbb{Z}}(\phi T_e^i W_e \cap T_o^j W_o)$, which implies (X_o, T_o) is a dissipative transformation, contradicting its being ergodic. This proves that the transformations $T_\mathbb{A}$ and $T_\mathbb{B}$ cannot be isomorphic. □

This completes the example. We also note that because of the definitions of the spaces X_o and X_e in $\mathbb{Z}_2$ it is easy to see that the sequences $\mathbb{A}$ and $\mathbb{B}$ are *eww* for T restricted to X_o and X_e respectively.

Proposition 7.6.7. *The eww sequences and sets for* T *restricted to* X_o *and* X_e
follows:

1. W_o *is an eww set for* T *on* $(X_o, \overline{\mu}_o)$ *for the sequence* $\mathbb{A} = SFS\{2^{2i+1} : i \geq 0\}$.
2. W_e *is an eww set for* T *on* $(X_e, \overline{\mu}_e)$ *for the sequence* $\mathbb{B} = SFS\{2^{2i} : i \geq 0\}$.

7.6.2 An Uncountable Family of Non-isomorphic Transformations

In this section we construct an uncountable family of non-isomorphic maps (see
[19]). We continue to work with the dyadic odometer T defined on the space $X \subset$
$\mathbb{Z}_2$ and again we define (as in the previous section) different measures on X.

Let us consider an infinite set of nonnegative integers $\mathbb{S} \subset \mathbb{N}$ such that $\mathbb{S}' = \mathbb{N} \backslash \mathbb{S}$
is also an infinite subset of $\mathbb{N}$. Associated to $\mathbb{S}$ we consider the following sets:

$$W_\mathbb{S} = \{x \in X : x_i = 0 \text{ for all } i \in \mathbb{S}\},$$
$$X_\mathbb{S} = \bigcup_{i \in \mathbb{Z}} T^i W_\mathbb{S},$$
$$\mathbb{A}_\mathbb{S} = SFS\{2^s : s \in \mathbb{S}\}.$$

We note that $W_\mathbb{S}$ "lives" on the coordinates of $\mathbb{N} \setminus \mathbb{S}$. On the two point set $\{0, 1\}$
we consider the two probability distributions $P = (1/2, 1/2)$ and $Q = (1, 0)$. We
define a product measure $\mu_\mathbb{S}$ on $W_\mathbb{S}$.

$$\mu_\mathbb{S} = \prod_{i=0}^{\infty} \eta_i, \quad \eta_i = \begin{cases} P & \text{if } i \in \mathbb{S}', \\ Q & \text{if } i \in \mathbb{S}. \end{cases}$$

With this measure the transformation induced on $W_\mathbb{S}$ by the first return map for the
odometer map T is again isomorphic to the von Neumann transformation and is
therefore ergodic.

As before we have the following lemmas.

Lemma 7.6.8. *The set* $X_\mathbb{S}$ *is invariant under* T *and satisfies the following:*

$$X_\mathbb{S} = \{x \in X : \text{ there exists an integer } k \geq 0, \text{ so that } x_s = 0 \text{ for all } s \geq k, \ s \in \mathbb{S}\}.$$

Lemma 7.6.9. *For distinct* $a, a' \in \mathbb{A}_\mathbb{S}$

$$T^a W_\mathbb{S} \cap T^{a'} W_\mathbb{S} = \emptyset.$$

Lemma 7.6.10. *The set* $X_\mathbb{S}$ *satisfies*

$$X_\mathbb{S} = \bigcup_{a \in \mathbb{A}_\mathbb{S}} T^a W_\mathbb{S} (disj).$$

The measure $\mu_{\mathbb{S}}$ on $W_{\mathbb{S}}$ can be pushed forward to all of $X_{\mathbb{S}}$:

$$\overline{\mu}_{\mathbb{S}}(E) = \sum_{s \in \mathbb{S}} \mu_{\mathbb{S}}(T^{-s}(E \cap T^{s} W_{\mathbb{S}})).$$

Proposition 7.6.11. *The map T is an infinite ergodic transformation on the measure space $(X_{\mathbb{S}}, \mu_{\mathbb{S}})$ and the sequence $\mathbb{A}_{\mathbb{S}}$ is an eww sequence with the eww set $W_{\mathbb{S}}$.*

There is an obvious duality between $\mathbb{S}$ and $\mathbb{S}'(= \mathbb{N} \setminus \mathbb{S})$ when both are infinite. We use this to construct our uncountable family of non-isomorphic transformations.

Let $\{\mathbb{S}_{\alpha}\}$ be an uncountable family of infinite subsets of $\mathbb{N}$ with pairwise finite intersections. It is well known how to construct such a family. Consider a 1-1 correspondence between the positive integers and the rationals. For each real number fix a sequence of rationals that converge to it. Then under the 1-1 correspondence, we obtain the desired family. We note that for every $\mathbb{S}_{\alpha}$ in this family the subset $\mathbb{S}'_{\alpha} = \mathbb{N} \setminus \mathbb{S}_{\alpha}$, is again an infinite subset of $\mathbb{N}$.

Now fix $\alpha \neq \beta$, and let $\mathbb{F} = \mathbb{S}_{\alpha} \cap \mathbb{S}_{\beta}$ denote the finite intersection of the sets $\mathbb{S}_{\alpha}$ and $\mathbb{S}_{\beta}$.

We let $W_{\alpha} = W_{\mathbb{S}'_{\alpha}} = \{x \in X : x_i = 0 \ \text{for all } i \in \mathbb{S}'_{\alpha}\}$ and $W_{\beta} = W_{\mathbb{S}'_{\beta}} = \{x \in X : x_i = 0 \text{ for all } i \in \mathbb{S}'_{\beta}\}$. We also let $V_{\alpha,\mathbb{F}} = \{x \in W_{\alpha} : x_i = 0 \ \text{for all } i \in \mathbb{F}\}$ and $V_{\beta,\mathbb{F}} = \{x \in W_{\beta} : x_i = 0 \ \text{for all } i \in \mathbb{F}\}$. Denote by μ_{α} the measure $\mu_{\mathbb{S}'_{\alpha}}$ constructed as above.

We let $X_{\alpha} = \cup_{i \in \mathbb{Z}} T^i W_{\alpha}$, and $X_{\beta} = \cup_{i \in \mathbb{Z}} T^i W_{\beta}$ as before; then both of these are invariant sets under T. It is also easy to see that $X_{\alpha} = \cup_{i \in \mathbb{Z}} T^i V_{\alpha,\mathbb{F}}$ and $X_{\beta} = \cup_{i \in \mathbb{Z}} T^i V_{\beta,\mathbb{F}}$.

Theorem 7.6.12. *For distinct indices $\alpha \neq \beta$, the infinite measure-preserving transformations $(X_{\alpha}, \mu_{\alpha}, T)$ and $(X_{\beta}, \mu_{\beta}, T)$ are non-isomorphic.*

Proof. The proof of Theorem 7.6.12 follows word for word the proof of Proposition 7.6.6 above once we use the next Lemma 7.6.13 to replace Lemma 7.6.5, replace the W sets in the proof of Proposition 7.6.6 by the corresponding V sets and make the appropriate subscript changes. We omit the details. □

Lemma 7.6.13. *For each $n \in \mathbb{N}$ either $T^n V_{\alpha,\mathbb{F}} \cap V_{\alpha,\mathbb{F}} = \emptyset$ or $T^n V_{\beta,\mathbb{F}} \cap V_{\beta,\mathbb{F}} = \emptyset$.*

Proof. We note that if $ord_2(n) \in \mathbb{S}'_{\alpha} \cup \mathbb{F}$ then

$$T^n V_{\alpha,\mathbb{F}} \cap V_{\alpha,\mathbb{F}} = \emptyset.$$

Similarly, if $ord_2(n) \in \mathbb{S}'_{\beta} \cup \mathbb{F}$ then

$$T^n V_{\beta,\mathbb{F}} \cap V_{\beta,\mathbb{F}} = \emptyset.$$

The lemma follows by noting that $(\mathbb{S}'_{\alpha} \cup \mathbb{F}) \cup (\mathbb{S}'_{\beta} \cup \mathbb{F}) = \mathbb{N}$. □

7.7 An Odometer Construction from $\mathcal{M}$

We have seen that the 2-adic integers are intimately connected to the pair of complementing sets in $\mathbb{N}$, $SFS\{2^{2i} : i = 0, 1, \ldots\}$ and $SFS\{2^{2i+1} : i = 0, 1, \ldots\}$. This complementing pair in $\mathbb{N}$ in turn came from a specific sequence of integers $\mathcal{M} = \{2, 2, 2, \ldots\}$, from the direct sum decompositions of $\mathbb{N}$ given by the de Bruijn and Vaidya structure theorem, Theorem 7.1.2. In this section we extend the results from the previous sections by starting from a general sequence of integers $\mathcal{M} = \{m_i : m_i \geq 2\}$ as given by the de Bruijn and Vaidya structure theorem. In Sect. 7.7.2, we will also construct the direct sum decomposition of $\mathbb{N}$ associated to $\mathcal{M}$ by the structure theorem; see [13].

7.7.1 Set Theoretic Construction of X

We assume that the sequence $\mathcal{M} = \{m_i : m_i \geq 2, \text{ for } i \geq 1\}$ has been fixed. We let

$$X_{\mathcal{M}} = \prod_{i=0}^{\infty} \{0, 1, \ldots, m_{i+1} - 1\},$$

and let T be the odometer map; that is

$$T : (x_0, x_1, x_2, \ldots, x_{n-1}, x_n, x_{n+1}, \ldots) = (0, 0, 0, \ldots, 0, x_n + 1, x_{n+1}, \ldots),$$

where $x_i = m_{i+1} - 1$ for $i < n$ and $x_n < m_{n+1} - 1$. This of course is the usual odometer action on the product space.

As before we treat the point $0 = (0, 0, 0, \ldots)$ separately. For the moment we remove from the space the points whose tail ends in all 0's or the tail ends with $(m_{i+1}-1)$'s for all $i > k$ and some k. This is a countable number of points removed. We will refer to this space as X. The proof of the following lemma is clear.

Lemma 7.7.1. *The map T on X is one-to-one and onto.*

Remark 7.7.2. It is well known that the space X may be viewed as a group where addition of two points is defined coordinate-wise with "carry to the right." In this sense the map T corresponds to "adding one" to the first coordinate, i.e.

$$T(x_0, x_1, x_2, \ldots) = (x_0, x_1, x_2, \ldots) + (1, 0, 0, \cdots),$$

with the sum having carry to the right.

We define the following sets in X—they are the analogs of the corresponding sets in Sect. 7.6:

$W_e = \{x \in X \mid x_{2i} = 0, \text{ for all } i\}$,
$W_o = \{x \in X \mid x_{2i+1} = 0, \text{ for all } i\}$,

$$X_o = \cup_{i \in \mathbb{Z}} T^i W_o,$$
$$X_e = \cup_{i \in \mathbb{Z}} T^i W_e.$$

Lemma 7.7.3. *The sets X_o and X_e are invariant under T and satisfy the following:*

1. $X_o = \{x \in X : \text{ for some integer } k \geq 0, \text{ so that } x_{2i+1} = 0 \text{ for all } i \geq k\}$,
2. $X_e = \{x \in X : \text{ for some integer } k \geq 0 \text{ such that } x_{2i} = 0 \text{ for all } i \geq k\}$.

Proof. Since we removed points whose tails ended in all 0's or $(m_{i+1} - 1)$'s adding (and subtracting) $(1, 0, 0, \ldots)$ to any point will only change a finite number of coordinates.   □

7.7.2   *Defining the Sequences* $\mathbb{A}$ *and* $\mathbb{B}$ *Associated to* $\mathcal{M}$

From the sequence $\mathcal{M} = \{m_i : m_i \geq 2, \text{ for } i \geq 1\}$ we will define a pair of complementing sequences $\mathbb{A}$ and $\mathbb{B}$ associated to the collection $\mathcal{M}$ (compare with the earlier $\mathbb{A}$ and $\mathbb{B}$ when all $m_i = 2$) as follows.

Let $M_0 = 1$ and define $M_k = \prod_{i=1}^{k} m_i$ for $k = 1, 2, \ldots$.

We define $\mathbb{A}$ to be the *SFS* set (Sums of Finite Subsets, see below) generated by the sequence

$$\{0, \underbrace{M_1, M_1, \ldots, M_1}_{m_2-1 \ times}, \underbrace{M_3, M_3, \ldots, M_3}_{m_4-1 \ times}, \ldots, \underbrace{M_{2n-1}, M_{2n-1}, \ldots, M_{2n-1}}_{m_{2n}-1 \ times}, \ldots\}$$

and we define $\mathbb{B}$ as the *SFS* set generated by the sequence

$$\{0, \underbrace{M_0, M_0, \ldots, M_0}_{m_1-1 \ times}, \underbrace{M_2, M_2, \ldots, M_2}_{m_3-1 \ times}, \ldots, \underbrace{M_{2n}, M_{2n}, \ldots, M_{2n}}_{m_{2n+1}-1 \ times}, \ldots\}.$$

Note that by the *SFS* set generated by a sequence $\{P_1, P_2, \ldots\}$ we mean the set of all finite sums $\{P_{i_1} + P_{i_2} + \cdots + P_{i_j}; 1 \leq i_1 < i_2 < \cdots < i_j, j = 1, 2, \ldots\}$. Compare this to the earlier use of the *SFS* of a sequence of integers, where we allowed the integers to be negative but all distinct. Here we require the integers to all be positive (except 0) but allow integers to repeat a finite number of times.

Remark 7.7.4. Each $a \in \mathbb{A}$ and $b \in \mathbb{B}$ may be viewed as members of $X_{\mathcal{M}}$ whose expansions end in a tail of 0's. Hence the action of T^a and T^b may also be viewed as addition by a and b respectively in the space X.

Remark 7.7.5. Observe that each $n \in \mathbb{N}$ is a finite sum of the form $n = \sum n_i M_i$ with $n_i \in \{0, 1, \ldots, m_{i+1} - 1\}$. Consequently, we can define $ord_{\mathcal{M}}(n) = i$ for the smallest i such that $n_i \neq 0$.

We define the sets $W_b = T^b W_e$ for $b \in \mathbb{B}$ and the sets $W_a = T^a W_o$ for $a \in \mathbb{A}$. Then we have the following:

Lemma 7.7.6. *The set W_e (W_o) is ww for $\mathbb{B}$ (respectively $\mathbb{A}$):*

1. *$W_b \cap W_{b'} = \emptyset$ for distinct $b \neq b'$ in $\mathbb{B}$; i.e. $T^b W_e \cap T^{b'} W_e = \emptyset$.*
2. *$W_a \cap W_{a'} = \emptyset$ for distinct $a \neq a'$ in $\mathbb{A}$; i.e. $T^a W_o \cap T^{a'} W_o = \emptyset$.*

Proof. Every positive integer n can be written as a finite sum $n = \sum n_i M_i$ where $n_i < m_{i+1}$. With the interpretation of T^n as addition by n we see that the representation of $a = \sum n_i M_i \in \mathbb{A}$ only has non-zero terms in odd coordinates and $b = \sum n_i M_i \in \mathbb{B}$ only has non-zero terms in even coordinates. Therefore T^b adds only to the "even" coordinates where X_e has 0's and so the two images $T^b W_e$ and $T^{b'} W_e$ are disjoint for $b \neq b' \in \mathbb{B}$. Similarly $T^a W_o \cap T^{a'} W_o$ are disjoint for distinct $a, a' \in \mathbb{A}$. $\square$

Remark 7.7.7. We note the following:

1. $X_e = \cup_{b \in \mathbb{B}} T^b W_e$ (disj).
2. $X_o = \cup_{a \in \mathbb{A}} T^a W_o$ (disj).

We also note that $X_o \cap X_e = \emptyset$ because we removed all the points with tails ending in 0 or ending in $m_{i+1} - 1$.

The remark above shows that $\mathbb{B}$ is an *eww* sequence on (X_e, T) with the set W_e; furthermore, $\mathbb{A}$ is an *eww* sequence for (X_o, T) with the set W_o. Of course we still need to define measures on X_e and X_o (see the proposition below).

Now we show that $\mathbb{A}$ has a distinguished role in X_e and $\mathbb{B}$ has a distinguished role in X_o. To explain this, put the point $\overline{0} = (0, 0, 0, \ldots)$ and all the points whose tail ends in all 0's back into the space X. This constitutes the forward orbit of the point $\overline{0}$. By definition of W_e some of these points will be in W_e and some will not. The set $\{n \geq 0 : T^n(0) \in W_e\}$ is our set $\mathbb{A}$. Furthermore, it is easy to see that the consecutive differences of $\{n \geq 0 : T^n(0) \in W_e\}$ generate the consecutive differences of almost all hitting times. Specifically, enumerate $\{n \geq 0 : T^n(0) \in W_e\} = \{0 = n_0 < n_1 < n_2 < n_3 < \cdots\}$. Define the consecutive difference sequence as $\{q_i : i \geq 1\}$ where $q_i = n_i - n_{i-1}$ for $i = 1, 2, \ldots$. Let x be an arbitrary point (in the *eww* set W_e) and let $\mathbb{H}(x) = \{\cdots < h_{-1} < 0 = h_0 < h_1 < \cdots\}$ be its hitting times to W_e. The consecutive difference sequence of $\mathbb{H}$ is $\{d_i = h_i - h_{i-1} : -\infty < i < \infty\}$.

Then for almost every point and every finite block $(d_i < d_{i+1} < \cdots < d_{i+k})$ of its consecutive difference sequence there is an $l \geq 1$ so that $(q_l < q_{l+1} < \cdots < q_{l+k}) = (d_i < d_{i+1} < \cdots < d_{i+k})$.

Of course such a point need not always exist, so we make the following definition of generic points in this setting as follows.

Definition 7.7.8. In this setting a one-sided sequence of integers $\mathbb{G} = \{0 = g_0 < g_1 < g_2 < \cdots\}$ will be said to be *generic* for the set A if for almost all points $x \in A$ the hitting sequence to A, $\mathbb{H}(x) = \{\cdots h_{-1} < 0 = h_0 < h_1 < \cdots\}$ satisfies the following two conditions:

- For all $n > 0$ the difference sequence $g_1 - g_0, g_2 - g_1, g_3 - g_2, \ldots, g_n - g_{n-1}$
  appears in $\mathbb{H}$ in the sense that there exists an i such that $g_k - g_{k-1} = h_{i+k} - h_{i+k-1}$ for $k = 1, \ldots n$.
- For all $n > 0$ the difference sequence $h_{-n+1} - h_{-n}, h_{-n+2} - h_{-n+1}, \ldots, h_n - h_{n-1}$
  appears in $\mathbb{G}$.

We now define a measure μ_e on X_e making the odometer action T an infinite ergodic transformation.

Proposition 7.7.9. *Suppose* $\mathbb{A}, \mathbb{B}$ *are constructed as above from the sequence* $\mathcal{M}$. *Then there is an infinite measure* μ_e *on* X_e, *so that action of* T *on* X_e *has the following properties:*

1. *The set* W_e *has measure one and is an eww set for* T *with the sequence* $\mathbb{B}$.
2. *The set* $\mathbb{A}$ *is generic for* W_e *in the sense of the definition above.*

Proof. We consider for each $i \geq 0$, the discrete spaces $\{0, 1, 2, \ldots, m_{i+1} - 1\}$ on which we define two probability measures P_i and Q_i where

$$P_i = (1/m_{i+1}, 1/m_{i+1}, \ldots, 1/m_{i+1}) \quad \text{and}$$

$$Q_i = (1, 0, 0, \ldots, 0).$$

On W_e we put the product measure $\prod_{i=0}^{\infty} \rho_i$ where $\rho_i = P_i$ for i odd and $\rho_i = Q_i$ for i even. Using T^b the measure on W_e is pushed forward to $W_b = T^b W_e$. We denote the sum measure on $X_e = \cup_{b \in \mathbb{B}} W_b$ by $\overline{\mu}_e$. $\qquad \square$

Likewise there is an infinite ergodic transformation on X_o for which the roles of $\mathbb{A}$ and $\mathbb{B}$ are interchanged. It is also possible to construct measures resulting in Type III transformations but we do not pursue that direction here.

As in the previous section the two transformations are non-isomorphic.

Theorem 7.7.10. *There is no isomorphism between the action of* T *on* $(X_o, \overline{\mu}_o)$ *and the action of* T *on* $(X_e, \overline{\mu}_e)$.

Proof. The proof is similar to the earlier case for $\mathcal{M}$ with $m_i = 2$ for all i. In particular, we have for all $n > 0$ either $T^n(W_e) \cap W_e = \emptyset$ or $T^n(W_o) \cap W_o = \emptyset$ and there cannot be an isomorphism. By Remark 7.7.5 we have that $ord_{\mathcal{M}}(n)$ is either odd or even. If it is odd then $T^n(W_o) \cap W_o = \emptyset$, and if it is even then $T^n(W_e) \cap W_e = \emptyset$. $\qquad \square$

7.7.3 The Sequence $\mathcal{M}$ and Multiple Recurrence

In this section we restrict our study to the transformation T on $(X_o, \overline{\mu}_o)$ constructed above; to simplify the notation, we will denote by μ the measure $\overline{\mu}_o$. We show how the parameters of the sequence $\mathcal{M}$ control the recurrence properties of the associated transformation T on (X_o, μ) (see [13]).

Definition 7.7.11. Let $p > 0$ be a positive integer. A transformation S defined on the measure space (X, ν) is *p-recurrent* if for any measurable set B with $\nu(B) > 0$ there is a positive integer $n > 0$ such that $\nu(B \cap S^n B \cap \cdots \cap S^{pn} B) > 0$. The transformation S is said to be *multiply recurrent* if S is p-recurrent for every $p > 0$.

We will prove the following:

Theorem 7.7.12. *Let T be the ergodic measure-preserving odometer on (X_o, μ) associated with the sequence $\mathscr{M} = \{m_i : m_i \geq 2 \text{ for all } i\}$. Then:*

1. T is multiply recurrent if and only if $\limsup(m_{2i}) = \infty$.
2. If $\limsup(m_{2i}) = p < \infty$ then T is $(p-1)$-recurrent but not p-recurrent.

We begin by defining the following class of cylinder sets. Let $(\varepsilon_0, \varepsilon_1, \ldots, \varepsilon_{2k-1})$ denote a set of $2k$ integers with $0 \leq \varepsilon_j < m_j$ for $0 \leq j < 2k$. The $2k$-*rectangle* is the subset

$$[\varepsilon_0, \varepsilon_1, \ldots, \varepsilon_{2k-1}] =$$

$$\{(x_0, x_1, \ldots) \in X_o : x_j = \varepsilon_j \text{ for } 0 \leq j < 2k \text{ and } x_{2i+1} = 0, \; \forall \, i \geq k\}.$$

It is clear that these $2k$-rectangular finite cylinder sets generate the σ-algebra. Therefore by the usual approximation arguments it is enough to prove recurrence for this class of sets.

The following lemmas show that we can calculate the intersections of a rectangle with its images.

Lemma 7.7.13. *Fix $K > 0$ and let R be a finite union of disjoint $2K$-rectangles. Then*

$$\mu(R \cap T^{M_{2k}} R \cap \cdots \cap T^{jM_{2k}} R) = (1 - j/m_{2k})\mu(R)$$

for $k \geq K$ and $0 < j < m_{2k}$.

Proof. Each $2K$-rectangle $[\varepsilon_0, \ldots, \varepsilon_{2K-1}]$ in R splits into m_{2K} $(2K + 2)$-sub-rectangles; namely $[\varepsilon_0, \ldots, \varepsilon_{2K-1}, \eta, 0]$ for $0 \leq \eta < m_{2K}$. Each one of these $(2K + 2)$-rectangles has measure $(1/m_{2K})\mu([\varepsilon_0, \ldots, \varepsilon_{2K-1}])$. Applying $T^{M_{2K}}$ we see exactly where each sub-rectangle maps. That is $T^{M_{2K}}[\varepsilon_0, \ldots, \varepsilon_{2K-1}, \eta, 0] = [\varepsilon_0, \ldots, \varepsilon_{2K-1}, \eta + 1, 0]$ for $0 \leq \eta < m_{2K} - 1$. Furthermore, the set $T^{M_{2K}}[\varepsilon_0, \ldots, \varepsilon_{2K-1}, m_{2K} - 1, 0] = [\varepsilon_0, \ldots, \varepsilon_{2K-1}, 0, 1]$ and thus is disjoint from all the initial $2K$-rectangles. The proof is identical for $k > K$. $\square$

Lemma 7.7.14. *Fix $K > 0$ and let R be a finite union of disjoint $2K$-rectangles. Then*

$$\mu(R \cap T^{jM_{2k}} R \cap T^{2jM_{2k}} R \cap \cdots \cap T^{m_{2k}jM_{2k}} R) = 0$$

for $k \geq K$ and $0 < j < m_{2k}$.

Proof. In this case we are taking the intersection of m_{2k} images. This is the same as adding jM_{2k} to the $2k - 1$ coordinate of the points in R. The values at the $2k - 1$ coordinate only takes the m_{2k} values $\{0, 1, \ldots, m_{2k} - 1\}$. Consequently, there must be a carry into the $2k$-th coordinate. As all the rectangles in R have the value 0 in this coordinate this results in an empty intersection.   □

Lemma 7.7.15. *Fix $K > 0$ and let R be a finite union of disjoint $2K$-rectangles. Then*

$$\mu(R \cap T^{jM_{2k+1}} R) = 0$$

for $k \geq K$ and $0 < j < m_{2k+1}$.

Proof. For all $k \geq K$, all points in the rectangles in R have a 0 in the $2k + 1$ coordinate. Applying $T^{jM_{2k+1}}$ with $0 < j < m_{2k+1}$, puts a j in the $2k + 1$ coordinate and is therefore disjoint from the initial rectangles in R.   □

Proof (Theorem 7.7.12). Suppose $\limsup(m_{2i}) = \infty$, and let B be a measurable set of positive measure. Then for any integer $p > 0$ there exist arbitrarily large even integers $2k$ such that $1 - p/m_{2k} \geq 1/p$. By possibly considering a subset of B we may assume that $0 < \mu(B) < \infty$. Since the $2k$-rectangles approximate the sets of finite measure it follows that there exists an even integer $2k > 0$ and a set R, which is the union of $2k$-rectangles in X_o, such that $\mu(R \Delta B) < (1/p)\mu(R)$. The previous lemmas then imply

$$\mu(R \cap T^{M_{2k}} R \cap \cdots \cap T^{(p-1)M_{2k}} R) = (1 - p/m_{2k})\mu(R) > (1/p)\mu(R).$$

It follows that $\mu(B \cap T^{M_{2k}} B \cap \cdots \cap T^{(p-1)M_{2k}} B) > 0$, and this shows that T is $(p - 1)$-recurrent for any $p > 0$.

Next suppose $p = \limsup(m_{2i}) < \infty$. Then there exists an integer $q > 0$, and we assume it to be even, such that $m_{2i} \leq p$ for all $2i > q$. From Lemma 7.7.15 when $\mathrm{ord}_{\mathcal{M}}(n) \geq q$, it follows that if $V = \{(x_0, x_1, \ldots) \in X_o : x_i = 0 \text{ for } 0 \leq i \leq q\}$ and $x \in V$, then there exists an integer j with $1 \leq j \leq p$ such that $T^{jn}x \notin V$. This implies that $V \cap T^n V \cap \cdots \cap T^{pn}V = \emptyset$, which says that the transformation T is not p-recurrent. In particular, T is not multiply recurrent.

Finally, we show that the transformation T is $(p - 1)$-recurrent in this case. We note that $p \geq 2$. Since $p = \limsup(m_{2i+1}) < \infty$ it follows that there exist arbitrarily large integers $2k > 0$ such that $m_{2k} = p$. Now let B be a set of positive measure. Again we may assume that $0 < \mu(B) < \infty$, and choose a large integer $2k > 0$ and a set R, which is the union of $2k$-rectangles in X_o, such that $\mu(R \Delta B) < (1/p)\mu(R)$. Recall that $M_{2k} = \prod_{1 \leq j \leq 2k} m_j$; Lemma 7.7.13 then implies $\mu(R \cap T^{M_i} R \cap \cdots \cap T^{(p-1)M_i} R) = (1/p)\mu(R)$. It follows that $\mu(B \cap T^{M_i} B \cap \cdots \cap T^{(p-1)M_i} B) > 0$, and this shows that T is $(p - 1)$-recurrent.   □

References

1. Aaronson, J.: An Introduction to Infinite Ergodic Theory. AMS Mathematical Surveys and Monographs, vol. 50, xii+284 pp. American Mathematical Society, Providence (1997)
2. Calderon, A.P.: Sur les mesures invariantes. C. R. Acad. Sci. Paris **240**, 1960–1962 (1955)
3. Clemens, J.D.: Descriptive Set Theory, Equivalence Relations, and Classification Problems in Analysis, 142 pp. Thesis (Ph.D.), University of California, Berkeley (2001)
4. Coven, E.M., Meyerowitz, A.: Tiling the integers with translates of one finite set. J. Algebra **212**, 161–174 (1999)
5. Dateyama, M., Kamae, T.: On direct sum decomposition of integers and Y. Ito's conjecture. Tokyo J. Math. **21**, 433–440 (1998)
6. de Bruijn, N.G.: On basis for the set of integers. Publ. Math. Debr. **1**, 232–242 (1950)
7. de Bruijn, N.G.: On number systems. Nieuw Arch. Wisk. (3) **4**, 15–17 (1956)
8. Dowker, Y.N.: On measurable transformations in finite measure spaces. Ann. Math. (2) **62**, 504–516 (1955)
9. Dowker, Y.N.: Sur les applications measurables. C. R. Acad. Sci. Paris **242**, 329–331 (1956)
10. Eigen, S., Hajian, A.: A characterization of exhaustive weakly wandering sequences for nonsingular transformations. Comment. Math. Univ. Sancti Pauli **36**, 227–233 (1987)
11. Eigen, S., Hajian, A.: Exhaustive weakly wandering sequences. Indag. Math. (N.S.) **18**, 527–538 (2007)
12. Eigen, S., Hajian, A.: Hereditary tiling sets of the integers. Integers **8**, A54, 9 pp. (2008)
13. Eigen, S., Hajian, A., Halverson, K.: Multiple recurrence and infinite measure preserving odometers. Isr. J. Math. **108**, 37–44 (1998)
14. Eigen, S., Hajian, A., Ito, Y.: Ergodic measure preserving transformations of finite type. Tokyo J. Math. **11**, 459–470 (1988)
15. Eigen, S., Hajian, A., Ito, Y., Prasad, V.: Existence and non-existence of a finite invariant measure, Tokyo J. Math. **35**, 339–358 (2012)
16. Eigen, S., Hajian, A., Kakutani, S.: Complementing sets of integers—a result from ergodic theory. Jpn. J. Math. (N.S.) **18**, 205–211 (1992)
17. Eigen, S., Hajian, A., Kalikow, S.: Ergodic transformations and sequences of integers. Isr. J. Math. **75**, 119–128 (1991)
18. Eigen, S., Hajian, A., Prasad, V.: Universal skyscraper templates for infinite measure preserving transformations. Discrete Continuous Dyn. Syst. **16**, 343–360 (2006)
19. Eigen, S., Hajian, A., Weiss, B.: Borel automorphisms with no finite invariant measure. Proc. Am. Math. Soc. **126**, 3619–3623 (1998)
20. Eigen, S., Ito, Y., Prasad, V.: Universally bad integers and the 2-adics. J. Number Theory **107**, 322–334 (2004)

© Springer Japan 2014

S. Eigen et al., *Weakly Wandering Sequences in Ergodic Theory*,

Springer Monographs in Mathematics, DOI 10.1007/978-4-431-55108-9

21. Friedman, N.: Introduction to Ergodic Theory. Van Nostrand Reinhold Mathematical Studies, vol. 29, v+143 pp. Van Nostrand Reinhold, New York (1970)
22. Fuglede, B.: Commuting self-adjoint partial differential operators and a group theoretic problem. J. Funct. Anal. **16**, 101–121 (1974)
23. Hajian, A.: Measurable Transformations and Invariant Measures. Thesis (Ph.D.), Yale University, New Haven (1957)
24. Hajian, A.: Strongly recurrent transformations. Pac. J. Math. **14**, 517–523 (1964)
25. Hajian, A.: On ergodic measure preserving transformations defined on an infinite measure space. Proc. Am. Math. Soc. **16**, 45–48 (1965)
26. Hajian, A., Ito, Y.: Iterates of measurable transformations and Markov operators. Trans. Am. Math. Soc. **17**, 371–386 (1965)
27. Hajian, A., Ito, Y.: Weakly wandering and related sequences. Z. Wahrscheinlichkeitstheorie Verw. Gebiete **8**, 315–324 (1967)
28. Hajian, A., Ito, Y.: Cesaro sums and measurable transformations. J. Comb. Theory **7**, 239–254 (1969)
29. Hajian, A., Ito, Y.: Weakly wandering sets and invariant measures for a group of transformations. J. Math. Mech. **18**, 1203–1216 (1969)
30. Hajian, A., Ito, Y.: Transformations that do not accept a finite invariant measue. Bull. Am. Math. Soc. **84**, 417–427 (1978)
31. Hajian, A., Ito, Y., Kakutani, S.: Invariant measures and orbits of dissipative transformations. Adv. Math. **9**, 52–65 (1972)
32. Hajian, A., Kakutani, S.: Weakly wandering sets and invariant measures. Trans. Am. Math. Soc. **110**, 136–151 (1964)
33. Hajian, A., Kakutani, S.: Example of an ergodic measure preserving transformation on an infinite measure space. In: 1970 Contributions to Ergodic Theory and Probability, pp. 45–52. Proceedings of the Conference, Ohio State University, Columbus. Springer, Berlin (1970)
34. Hajós, G.: Sur la factorisation des groupes abéliens. Časopis Pěst. Mat. Fys. **74**, 157–162 (1949)
35. Hamachi, T., Osikawa, M.: On zero type and positive type transformations with infinite invariant measures. Mem. Fac. Sci. Kyushu Univ. Ser. A **25**, 280–295 (1971)
36. Hansen, R.T.: Complementing pairs of subsets of the plane. Duke Math. J. **36**, 441–449 (1969)
37. Hopf, E.: Theory of measures and invariant integrals. Trans. Am. Math. Soc. **34**, 353–373 (1932)
38. Hopf, E.: Ergodentheorie, v+83 pp. Springer, Berlin (1937)
39. Ito, Y.: Invariant Measures for Markov Processes. Thesis (Ph.D.), Yale University, New Haven (1962)
40. Ito, Y.: Invariant measures for Markov processes. Trans. Am. Math. Soc. **110**, 152–184 (1964)
41. Ito, Y.: Direct sum decomposition of the integers. Tokyo J. Math. **18**, 259–270 (1995)
42. Jones, L., Krengel, U.: On transformations without finite invariant measure. Adv. Math. **12**, 275–295 (1974)
43. Kakutani, S.: Induced measure preserving transformations. Proc. Imp. Acad. Tokyo. **30**, 635–641 (1943)
44. Kakutani, S.: Classification of ergodic groups of automorphisms. In: Proceedings of the International Conference on Functional Analysis and Related Topics, pp. 392–397, Tokyo, April 1969
45. Kakutani, S.: A Problem of Equidistribution on the Unit Interval. Lecture Notes in Mathematics, vol. 541, pp. 369–375. Springer, Berlin (1976)
46. Kamae, T.: A characterization of weakly wandering sequences for nonsingular transformations. Comment. Math. Univ. St. Paul. **32**, 55–59 (1983)
47. Kosek, W.: Dissipative sequences in infinite ergodic theory. Adv. Math. **180**, 427–451 (2003)
48. Long, C.T.: Addition theorems for sets of integers. Pac. J. Math. **23**, 107–112 (1967)
49. Newman, D.J.: Tesselation of integers. J. Number Theory **9**, 107–111 (1977)
50. Niven, I.: A characterization of complementing sets of pairs of integers. Duke Math. J. **38**, 193–203 (1971)

51. Ornstein, D.S., Shields, P.C.: An uncountable family of K-automorphisms. Adv. Math. **10**, 63–88 (1973)
52. Swenson, C.: Direct sum subset decompositions of $\mathbb{Z}$. Pac. J. Math. **53**, 629–633 (1974)
53. Vaidya, A.M.: On complementing sets of nonnegative integers. Math. Mag. **39**, 43–44 (1966)
54. Vershik, A.: Uniform algebraic approximation of shift and multiplicative operators. Dokl. Akad. Nauk SSSR **218**, 526–529 (1981)

Index

A
Aaronson, J., 94, 102
absolutely continuous, 6
adding machine, 43, 48, 52, 90
α-type, 86, 87, 93

B
Birkhoff Ergodic Theorem, x, 26

C
Calderon, A.P., 5, 34
Clemens, J.D., 115–117
commutator, 49, 69, 81
complementing pair, 104–106, 108, 109
complementing set, 104, 116, 119, 132
compressible set, 10
countably equivalent sets, 2, 5, 10, 14, 15, 27
Coven, E., 104–106
cutting and stacking, 84, 85, 88, 96, 111, 115, 119

D
Dateyama, M., 132
de Bruijn, N.G., 104, 105, 141
derived set, 17, 18
difference set, 65, 66, 110, 116, 117
direct sum, 103, 106, 108–110
dissipative sequence, 71, 72, 74, 75
dissipative transformation, 49, 57
Dowker, Y., ix, xi, 5, 34
dyadic odometer, 43, 48, 84, 88, 90, 102

E
Eigen, S., 23, 67, 95, 102, 106, 107, 111, 115, 119, 124, 131, 132, 139, 141, 145
equivalent measures, 1
ergodic, 21, 25
ergodic type III, 49, 50
even complete, 132
even differences, 131, 132
eww, see exhaustive weakly wandering
eww growth sequence, 36, 37, 39
eww sequence, 18, 23, 36, 39, 46, 67, 85, 96, 108–112, 116, 117, 119, 123, 140, 143
eww set, 46, 48, 67, 69, 85, 88, 96, 109, 112, 123, 140, 143
exh, see exhaustive
exh sequence, 67
exh set, 67
exhaustive, 67
exhaustive weakly wandering, 18
exterior, 76

F
finite ergodic, 25
finite invariant measure, ix, 1, 4–16, 29
finite tilings, 105, 106
finite type, 67, 80
finitely equivalent sets, 2, 3, 8, 10
foundational sequence, 92
free ultrafilter, 75
Friedman, N., 94, 111, 112
Fuglede conjecture, 106
Fuglede, B., 106
full group, 53

© Springer Japan 2014
S. Eigen et al., *Weakly Wandering Sequences in Ergodic Theory*,
Springer Monographs in Mathematics, DOI 10.1007/978-4-431-55108-9

Printed in the United States
By Bookmasters

Printed in the United States
By Bookmasters